AF594936

Driving Sustainability through Engineering Management and Systems Engineering

Driving Sustainability through Engineering Management and Systems Engineering

Editor

Simon P. Philbin

MDPI • Basel • Beijing • Wuhan • Barcelona • Belgrade • Manchester • Tokyo • Cluj • Tianjin

Editor
Simon P. Philbin
Nathu Puri Institute for
Engineering and Enterprise
London South Bank University
London
United Kingdom

Editorial Office
MDPI
St. Alban-Anlage 66
4052 Basel, Switzerland

This is a reprint of articles from the Special Issue published online in the open access journal *Sustainability* (ISSN 2071-1050) (available at: www.mdpi.com/journal/sustainability/special_issues/En_management).

For citation purposes, cite each article independently as indicated on the article page online and as indicated below:

LastName, A.A.; LastName, B.B.; LastName, C.C. Article Title. *Journal Name* **Year**, *Volume Number*, Page Range.

ISBN 978-3-0365-1532-8 (Hbk)
ISBN 978-3-0365-1531-1 (PDF)

Contents

About the Editor

Simon P. Philbin

Simon P. Philbin is Professor and Director of Engineering and Enterprise in the School of Engineering at London South Bank University (LSBU) in the United Kingdom, where he currently leads the Nathu Puri Institute for Engineering and Enterprise. His primary research interests are in the area of engineering management with a particular focus on sustainability and technology-driven innovation. He holds a PhD (Brunel University London) and BSc (University of Birmingham), both in chemistry, as well as an MBA (Open University Business School). He has authored/co-authored over 75 peer-reviewed papers and has presented research studies at various international conferences across North America and Asia. He served as the 2019/20 President of the American Society for Engineering Management. Previous research positions include Visiting Fellow at Imperial College Business School and Visiting Fellow in the School of Business, Economics and Informatics at Birkbeck, University of London.

Preface to "Driving Sustainability through Engineering Management and Systems Engineering"

As the world continues to respond to the COVID-19 pandemic, there is a pressing need to address the corresponding health and economic crises that have gripped many nations. The continued deployment of vaccines is targeted towards alleviating the health situation, and both governments and central banks combined with the private sector are focused on enabling economic recovery. Clearly, these matters are of paramount importance. However, the need to ensure we transition to a more sustainable development path remains a priority; although it may have been somewhat overshadowed by recent events, it certainly has not gone away. Indeed, while some reports indicated that carbon dioxide emissions were reduced in the first half of 2020, they rapidly picked up to the previously unsustainable levels in the latter half of the year. Tackling climate change and global warming remains a major challenge towards the top of the list of priorities on the journey to sustainability, although there are of course many others that are part of the United Nations (UN) Sustainable Development Goals (SDGs).

If we are to heed the warnings of the UN Intergovernmental Panel on Climate Change (IPCC) and keep the global temperature increase to within 1.5°C, there is an urgent need for action across multiple fronts. This includes a continued focus across the triple bottom line of economic, environmental and social requirements combined with leveraging the latest scientific and technological developments. Moreover, while the challenge of green growth and sustainable development continues, it is accompanied by the parallel opportunity of harnessing the latest technologies. This includes emerging technologies associated with Industry 4.0 as well as others, such as developments in the renewable energy sector. Therefore, it is important that we have the available tools and techniques to tackle the twin challenges and opportunities associated with sustainability.

The disciplines of engineering management and systems engineering offer such a toolset. Engineering management is concerned with the management of people and projects related to technological and engineering systems, while systems engineering is focused on the design, integration and management of complex systems across the full development life cycle. Engineering management and systems engineering provide both a theoretical foundation and complementary set of practitioner-oriented tools, techniques and models that can be used to develop and manage across the project, programme, portfolio and organisational dimensions. This includes both the technological and people dimensions associated with system design and technology management applications. Consequently, engineering management and systems engineering are ideally suited to drive sustainability.

This book includes chapters that reveal new insights into sustainability according to various perspectives. The chapters comprise a range of empirical and conceptual methods that enable quantitative and qualitative investigations of the underlying phenomenon of interest. Although the chapters adopt different approaches from across engineering management and systems engineering, they have a common focus on advancing our understanding of how sustainable development can be supported and delivered. The chapters provide details on a range of different research studies, and this includes understanding how to measure the SDG impact of infrastructure projects; development of an improved process for selecting sustainable projects; evaluation of the impact of COVID-19 on the food and beverages manufacturing sector; understanding the adoption of complexity theory in engineering management and the implications for sustainability; investigating the scope for a vehicle-to-grid strategy to improve the energy performance of buildings; and understanding how the data centre industry can be transformed into a circular economy model. In conclusion, sustainable development and achievement of the SDGs represent an important journey, where the disciplines of engineering management and systems engineering can be viewed as the vehicles to ensure the journey towards sustainability is continued.

Simon P. Philbin
Editor

Editorial

Driving Sustainability through Engineering Management and Systems Engineering

Simon P. Philbin 

Nathu Puri Institute for Engineering and Enterprise, School of Engineering, London South Bank University, 103 Borough Road, London SE1 0AA, UK; philbins@lsbu.ac.uk

Citation: Philbin, S.P. Driving Sustainability through Engineering Management and Systems Engineering. *Sustainability* **2021**, *13*, 6687. https://doi.org/10.3390/su13126687

Received: 10 June 2021
Accepted: 11 June 2021
Published: 12 June 2021

Publisher's Note: MDPI stays neutral with regard to jurisdictional claims in published maps and institutional affiliations.

The COVID-19 pandemic has highlighted the somewhat precarious nature of our lives, including the way we work and our lifestyles. It is clearly important that efforts continue to be directed towards remedying this situation; and thankfully the global rollout of vaccines is beginning to have a positive impact although there is of course still much to be done across the world [1]. In this context, however, the seemingly ever-present challenge of realizing sustainability across the triple bottom line of social, environmental, and economic development has not diminished [2–4]. If anything, it is now more pressing than ever that we work towards achieving the United Nations' Sustainable Development Goals (SDGs) [5,6]. The challenges appear to be significant both in scale and complexity.

The burning of hydrocarbons continues apace, leading to climate change and the resulting negative consequences [7]. A reduction in carbon emissions [8] and decarbonization of industrial supply chains [9] is required along with concomitant adoption of renewable energy production, such as solar and wind power [10]. Despite advances in such areas, there remain various issues associated with the adoption of renewable forms of energy, such as solar power, including the low capacity factor, grid instability, and intermittency [11]. Furthermore, there is a need to ensure effective energy storage solutions are available to complement solar power provision [12], a need to understand the environmental risks of solar cells [13], and a need to move towards a circular economy use of photovoltaics [14].

The level of urbanization continues to increase along with the growing ecological footprint and decreasing biocapacity, which represent a major challenge to be addressed in many parts of the world [15]. Environmental pollution caused by high levels of industrial development has resulted in degradation of the quality of freshwater along with depletion of this finite resource [16]. There remain serious issues associated with alleviating the multi-dimensional nature of poverty in developing nations [17,18]. The list of challenges to sustainability goes on. Many of these societal challenges represent complex problems concerning technical, organisational, social, and political issues, where a paradigm shift is required in order to understand and engage with such issues [19]. Such a paradigm shift needs to be informed by an appropriate form of thinking and knowledge creation, including the capture, codification, and analysis of relevant data and information. In this regard and from an epistemological perspective, the disciplines of engineering management and systems engineering provide a viable and robust solution to help tackle the challenges of sustainable development and the broader needs for sustainability.

Engineering management involves the management of people and projects related to technological or engineering systems—this includes project management, engineering economy, and technology management, as well as the management and leadership of teams. The American Society for Engineering Management (ASEM) defines engineering management as "an art and science of planning, organizing, allocating resources, and directing and controlling activities that have a technological component" [20]. Engineering management is used across different industrial sectors as part of the management of engineering programmes and technology solutions [21]. With regard to research studies, the field of engineering management spans different areas, for instance, ecological engineering [22], operational design [23], safety engineering [24], and lean construction projects [25].

Systems engineering involves the design, integration, and management of complex systems over the full life cycle—this includes requirements capture, integrated system design, systems integration, as well as modelling and simulation. The International Council on Systems Engineering [26] defines systems engineering as "an interdisciplinary approach and means to enable the realization of successful systems. It focuses on holistically and concurrently understanding stakeholder needs; exploring opportunities; documenting requirements; and synthesizing, verifying, validating, and evolving solutions while considering the complete problem, from system concept exploration through system disposal". Systems engineering is utilised in different applications and across different industrial settings to help in the design, development, and implementation of complex industrial and organisational systems [27]. With regard to research studies, the field of systems engineering spans different areas, for instance, chemical process systems engineering [28], satellite systems [29], vehicle system design [30], and renewable energy systems [31].

In addition to the theoretical underpinnings of both disciplines, they also provide a range of tools and techniques that can be employed to address complexity across technological, organizational, social, and political domains. Such tools and techniques include, for instance, project management, financial analysis, R&D management, operations management, requirements capture, engineering design and modelling, optimization methods, workflow analysis, and system dynamics. The disciplines of engineering management and systems engineering are therefore ideally suited to help tackle both the challenges and opportunities associated with realizing a sustainable future. Furthermore, engineering management and systems engineering implicitly enable multifaceted solutions informed by multidisciplinary approaches in order to provide the necessary knowledge and processes to underpin sustainability.

As mentioned, the goal of sustainable development presents many challenges. However, there are also various opportunities that are now becoming available. Indeed, the 4th Industrial Revolution, or Industry 4.0, is gathering pace across manufacturing [32–34] along with there being significant scope for adoption in other sectors, such as the energy and food sectors [35]. Innovation 4.0 can be regarded as a collection of different but related technologies that enable integration between physical and digital systems. The related technologies include cyber-physical systems; the industrial internet-of-things; artificial intelligence and machine learning; autonomous robots; big data and big data analytics; simulation; digital twins; virtual and augmented reality; cloud computing; cybersecurity; and additive manufacturing (3D-printing). Looking ahead, it is likely that adoption of 5G wireless technology for digital cellular networks along with much faster download speeds will help to power ahead adoption of Industry 4.0 technologies across new applications [36].

Not only are the technologies and corresponding new business processes from the Industry 4.0 paradigm positioned to support improvements in productivity as well as industrial competitiveness [37] but there are also tangible links to enabling energy sustainability, harmful emission reduction, and social welfare improvement [38]. While still an emerging area, there are already studies examining the link between Industry 4.0 and sustainability, including related aspects such as the circular economy, resource scarcity, lean production, energy problems, food production, process management and modelling, waste reduction, raw material reuse, and social aspects [39]. Furthermore, Industry 4.0 technologies have the capacity to improve the sustainability of supply chains [40] as well as support the adoption of sustainable business models [41].

Across this aforementioned context, this Special Issue in Sustainability draws on the theory and practical application of engineering management and systems engineering in order to drive sustainability. The contributions in the Special Issue address different perspectives and approaches that enable sustainability through leveraging the disciplines of engineering management and systems engineering. The scope of research studies includes measuring the sustainability of infrastructure projects [42]; sustainable project selection [43]; assessing the impact of COVID-19 on the food and beverages manufacturing sector [44]; complexity theory in engineering management [45]; vehicle to grid technology for improved energy performance [46]; and development of a data centre industry towards a circular economy [47].

The contributions include the use of different research methodologies in order to investigate the underlying phenomenon of interest and this includes qualitative research via expert interviews [42]; quantitative modelling via the fuzzy analytic hierarchy process [43]; quantitative research via use of a survey instrument [44]; conceptual studies through analysis of extant literature [45]; engineering modelling based on machine learning [46]; and a mixed quantitative/qualitative method based on a survey instrument and semi-structured interviews [47]. The Special Issue also has international contributions representing a global coverage of the subject matter, including three articles from authors in the United Kingdom, one article from Italy, one article from the United States, and one article from South Africa.

The first contribution is by Mansell et al. [42] on "Delivering UN Sustainable Development Goals' Impact on Infrastructure Projects: An Empirical Study of Senior Executives in the UK Construction Sector". This research examined how the United Nations SDGs can be measured on infrastructure projects through a qualitative study involving interviews with 40 executives from the United Kingdom (UK) construction industry. Although the SDGs carry a certain level of complexity through there being 17 global goals along with 169 targets and 244 indicators, the study did however identify both the opportunities and challenges for measuring the SDGs in this application. This includes a recognition that measuring SDG performance should include a full project lifecycle perspective as well as an adequate understanding of the longer-term project outcomes and wider impacts generated by infrastructure projects. Moreover, there is merit in viewing infrastructure projects from a systems perspective in order to fully capture the wider impact, including the link to sustainability metrics. The results further indicate that while SDG measurement practices are embraced in principle, there remain difficulties in how they can be implemented. Nevertheless, there continues to be significant scope for those involved in the management of infrastructure projects to apply the findings of the study in order to translate the SDGs to the project level.

The second contribution is by Alyamani and Long [43] on "The Application of Fuzzy Analytic Hierarchy Process in Sustainable Project Selection". This research investigated the process of sustainable project selection as part of enabling sustainable development through achievement of project goals. Since the process of sustainable project selection depends on a wide range of factors, there can be difficulties associated with the subjective judgments of experts involved in prioritizing project selection criteria as well as the resulting uncertainties associated with the subjective judgments by such experts. The study employed the fuzzy analytic hierarchy process (FAHP) as the mechanism to account for these uncertainties, where 'expert opinions' were gathered from the literature and translated into the necessary variables. Specifically, FAHP was used to rank five key sustainable project selection criteria identified by the study through calculating the relative weight of importance for each of the selection criteria. The results from the study identified that the most important criterion to consider during the process of sustainable project selection is project cost, followed by novelty and uncertainty as the second and third most important criteria, respectively.

The third contribution is by Telukdarie et al. [44] on "Analysis of the Impact of COVID-19 on the Food and Beverages Manufacturing Sector". This research examined how the food and beverages manufacturing sector in South Africa can respond to the major challenges caused by the COVID-19 global pandemic. The study considered a number of different sources of information from the global literature as part of a knowledge classification process in order to formulate an expedited response by the food and beverages manufacturing sector. This enabled deployment of a survey instrument with 106 small and medium-sized enterprises (SMEs) engaged in the food and beverages manufacturing sector in South Africa. Following statistical analysis via use of Cronbach Alpha to test the reliability of the data, the results identified that SMEs are under significant pressure caused by the pandemic. The results further revealed that the situation can be mitigated through a number of different measures, including social distancing, communication, facilities reconfiguration, and virtual working, which aligns well with international best practice. Additionally, the study pointed towards further mitigation measures being adopted beyond South Africa and globally, and this includes human resource and workforce adjustments as well as scope to leverage technology developments associated with the Industry 4.0 paradigm.

The fourth contribution is by Abatecola and Surace [45] on "Discussing the Use of Complexity Theory in Engineering Management: Implications for Sustainability". This research explored through a conceptual lens the adoption of complexity theory (CT) in the field of engineering management as well as the resulting implications concerning sustainability. The current status of complexity research in EM was identified and discussed through review of 38 sources of relevant literature published in the journal, IEEE Transactions on Engineering Management, which provided a heuristic proxy as part of the conceptual study. The research identified that since the year 2000, CT in EM has been associated with a wide range of key themes in the field, namely, new product development, supply chain management, and project management. Moreover, the dataset included modelling and optimizing decision-making under uncertainty as a key theme related to CT in the EM literature. In terms of the link to the area of sustainability, the research identified that there is scope to apply the use of CT to this application in areas such as energy, healthcare, and construction, which could be supported through harnessing complexity-based observations. The study also proposed that firms engaged in tackling sustainability issues would benefit from adopting structures and processes according to a complex adaptive system model.

The fifth contribution is by Scott et al. [46] on "Machine Learning Based Vehicle to Grid Strategy for Improving the Energy Performance of Public Buildings". This research investigated vehicle-to-grid (V2G) technology as a way for carbon neutral buildings to accommodate situations involving unpredictable renewable sources. The study provides an effective V2G strategy, which was developed and implemented for an operational university campus. Furthermore, the study derived a machine learning algorithm in order to predict energy consumption and energy costs for the building on the university campus. This approach was based on a feed-forward neural network (FFNN) machine learning algorithm (MLA) that enabled prediction of the future energy demand for the building under investigation, where the input data was taken from 2013–2017 to predict output data for the years 2018 and 2019. The research focused on integrating V2G technology into the campus environment according to different scenarios and various demand levels, such as peak time and off-peak, and included modelling of net profits. The engineering modelling study revealed that the proposed V2G installation at the university campus is both economically and environmentally beneficial, thereby underpinning the case for sustainability of this technology solution.

The sixth contribution is by Andrews et al. [47] on "A Circular Economy for the Data Centre Industry: Using Design Methods to Address the Challenge of Whole System Sustainability in a Unique Industrial Sector". This research explored how the data centre industry (DCI) can be transformed from a linear economic model to a circular economy system. As a major global industry, there are significant challenges for current practices in the DCI, which are becoming both environmentally and socially unsustainable. Therefore, a new approach to enable transformation to a circular economy model has been developed that is based on a whole systems perspective, design thinking, and the double diamond method, combined with people/stakeholder engagement throughout the project. A whole systems approach has benefits in terms of capturing how decisions are made and actions are taken at each stage in the life cycle, which can impact the sustainability of the remaining stages in the life cycle. The study collected quantitative data via an online survey along with deeper insights through qualitative research from semi-structured interviews. The findings revealed that the project under investigation is generating innovative outputs (namely, designs, business models, and a digital tool), which can support the DCI to follow a circular economy trajectory.

The published research articles in the Special Issue are as follows:

1. Delivering UN Sustainable Development Goals' Impact on Infrastructure Projects: An Empirical Study of Senior Executives in the UK Construction Sector by Paul Mansell, Simon P. Philbin, and Efrosyni Konstantinou [42].
2. The Application of Fuzzy Analytic Hierarchy Process in Sustainable Project Selection by Rakan Alyamani and Suzanna Long [43].
3. Analysis of the Impact of COVID-19 on the Food and Beverages Manufacturing Sector by Arnesh Telukdarie, Megashnee Munsamy, and Popopo Mohlala [44].
4. Discussing the Use of Complexity Theory in Engineering Management: Implications for Sustainability by Gianpaolo Abatecola and Alberto Surace [45].
5. Machine Learning Based Vehicle to Grid Strategy for Improving the Energy Performance of Public Buildings by Connor Scott, Mominul Ahsan, and Alhussein Albarbar [46].
6. A Circular Economy for the Data Centre Industry: Using Design Methods to Address the Challenge of Whole System Sustainability in a Unique Industrial Sector by Deborah Andrews, Elizabeth J. Newton, Naeem Adibi, Julie Chenadec, and Katrin Bienge [47].

Funding: This work was not supported by any specific funding.

Conflicts of Interest: The author declares that there is no conflict of interest.

References

1. Kim, J.H.; Marks, F.; Clemens, J.D. Looking beyond COVID-19 vaccine phase 3 trials. *Nat. Med.* **2021**, *27*, 205–211. Available online: https://www.nature.com/articles/s41591-021-01230-y (accessed on 9 June 2021). [CrossRef]
2. Neri, A.; Cagno, E.; Lepri, M.; Trianni, A. A triple bottom line balanced set of key performance indicators to measure the sustainability performance of industrial supply chains. *Sustain. Prod. Consum.* **2021**, *26*, 648–691. Available online: https://www.sciencedirect.com/science/article/abs/pii/S2352550920314172 (accessed on 9 June 2021). [CrossRef]
3. Ecer, F. Sustainability assessment of existing onshore wind plants in the context of triple bottom line: A best-worst method (BWM) based MCDM framework. *Environ. Sci. Pollut. Res.* **2021**, *28*, 19677–19693. [CrossRef]
4. Raza, F.; Alshameri, B.; Jamil, S.M. Assessment of triple bottom line of sustainability for geotechnical projects. *Environ. Dev. Sustain.* **2021**, *23*, 4521–4558. [CrossRef]
5. Nhamo, G.; Chikodzi, D.; Kunene, H.P.; Mashula, N. COVID-19 vaccines and treatments nationalism: Challenges for low-income countries and the attainment of the SDGs. *Glob. Public Health* **2021**, *16*, 319–339. [CrossRef] [PubMed]
6. Berger, M.; Campos, J.; Carolli, M.; Dantas, I.; Forin, S.; Kosatica, E.; Kramer, A.; Mikosch, N.; Nouri, H.; Schlattmann, A.; et al. Advancing the water footprint into an instrument to support achieving the SDGs–recommendations from the "Water as a Global Resources" research initiative (GRoW). *Water Resour. Manag.* **2021**, *35*, 1291–1298. [CrossRef]
7. Hornsey, M.J.; Fielding, K.S. Understanding (and reducing) inaction on climate change. *Soc. Issues Policy Rev.* **2020**, *14*, 3–35. [CrossRef]
8. Adams, S.; Acheampong, A.O. Reducing carbon emissions: The role of renewable energy and democracy. *J. Clean. Prod.* **2019**, *240*, 118245. [CrossRef]
9. Kazi, M.K.; Eljack, F.; El-Halwagi, M.M.; Haouari, M. Green hydrogen for industrial sector decarbonization: Costs and impacts on hydrogen economy in Qatar. *Comput. Chem. Eng.* **2021**, *145*, 107144. Available online: https://www.sciencedirect.com/science/article/abs/pii/S009813542030836X (accessed on 9 June 2021). [CrossRef]
10. Weschenfelder, F.; Leite, G.D.N.P.; da Costa, A.C.A.; de Castro Vilela, O.; Ribeiro, C.M.; Ochoa, A.A.V.; Araujo, A.M. A review on the complementarity between grid-connected solar and wind power systems. *J. Clean. Prod.* **2020**, *257*, 120617. Available online: https://www.sciencedirect.com/science/article/abs/pii/S0959652620306648 (accessed on 9 June 2021). [CrossRef]
11. Li, K.; Liu, C.; Jiang, S.; Chen, Y. Review on hybrid geothermal and solar power systems. *J. Clean. Prod.* **2020**, *250*, 119481. Available online: https://www.sciencedirect.com/science/article/abs/pii/S0959652619343513 (accessed on 9 June 2021). [CrossRef]
12. Guerra, O.J.; Zhang, J.; Eichman, J.; Denholm, P.; Kurtz, J.; Hodge, B.M. The value of seasonal energy storage technologies for the integration of wind and solar power. *Energy Environ. Sci.* **2020**, *13*, 1909–1922. Available online: https://pubs.rsc.org/am/content/articlelanding/2020/ee/d0ee00771d/unauth#!divAbstract (accessed on 9 June 2021). [CrossRef]
13. Kwak, J.I.; Nam, S.H.; Kim, L.; An, Y.J. Potential environmental risk of solar cells: Current knowledge and future challenges. *J. Hazard. Mater.* **2020**, *392*, 122297. Available online: https://www.sciencedirect.com/science/article/abs/pii/S0304389420302855 (accessed on 9 June 2021). [CrossRef]

14. Heath, G.A.; Silverman, T.J.; Kempe, M.; Deceglie, M.; Ravikumar, D.; Remo, T.; Cui, H.; Sinha, P.; Libby, C.; Shaw, S.; et al. Research and development priorities for silicon photovoltaic module recycling to support a circular economy. *Nat. Energy* **2020**, *5*, 502–510. Available online: https://www.nature.com/articles/s41560-020-0645-2 (accessed on 9 June 2021). [CrossRef]
15. Ahmed, Z.; Zafar, M.W.; Ali, S. Linking urbanization, human capital, and the ecological footprint in G7 countries: An empirical analysis. *Sustain. Cities Soc.* **2020**, *55*, 102064. Available online: https://www.sciencedirect.com/science/article/abs/pii/S2210670720300512 (accessed on 9 June 2021). [CrossRef]
16. Qadri, H.; Bhat, R.A. The concerns for global sustainability of freshwater ecosystems. In *Fresh Water Pollution Dynamics and Remediation*; Springer: Singapore, 2020. [CrossRef]
17. Singh, P.K.; Chudasama, H. Evaluating poverty alleviation strategies in a developing country. *PLoS ONE* **2020**, *15*, e0227176. [CrossRef] [PubMed]
18. Yang, Y.; de Sherbinin, A.; Liu, Y. China's poverty alleviation resettlement: Progress, problems and solutions. *Habitat Int.* **2020**, *98*, 102135. Available online: https://www.sciencedirect.com/science/article/abs/pii/S0197397519303960 (accessed on 9 June 2021). [CrossRef]
19. Da Costa Junior, J.; Diehl, J.C.; Snelders, D. A framework for a systems design approach to complex societal problems. *Des. Sci.* **2019**, *5*, e2. Available online: https://www.cambridge.org/core/journals/design-science/article/framework-for-a-systems-design-approach-to-complex-societal-problems/058568678B9FF25E6653EE53925838F8 (accessed on 9 June 2021). [CrossRef]
20. ASEM. *A Guide to the Engineering Management Body of Knowledge*, 5th ed.; Hiral, S., Walter, N., Eds.; American Society for Engineering Management (ASEM): Huntsville, AL, USA, 2019; ISBN 978-0-9975195-5-6. Available online: https://www.asem.org/EMBoK (accessed on 9 June 2021).
21. Philbin, S.P.; Kennedy, D. Exploring the need for a new paradigm in engineering management and the decision-making process in technology-based organisations. *Eng. Manag. Prod. Serv.* **2020**, *12*, 7–21. Available online: https://sciendo.com/de/article/10.2478/emj-2020-0024 (accessed on 9 June 2021).
22. Xu, J.; Li, Z. A review on ecological engineering based engineering management. *Omega* **2012**, *40*, 368–378. Available online: https://www.sciencedirect.com/science/article/abs/pii/S0305048311000995 (accessed on 9 June 2021). [CrossRef]
23. Coates, G.; Duffy, A.H.; Whitfield, I.; Hills, W. Engineering management: Operational design coordination. *J. Eng. Des.* **2004**, *15*, 433–446. [CrossRef]
24. Philbin, S.P. Developing an integrated approach to system safety engineering. *Eng. Manag. J.* **2010**, *22*, 56–67. [CrossRef]
25. Bajjou, M.S.; Chafi, A. Identifying and managing critical waste factors for lean construction projects. *Eng. Manag. J.* **2020**, *32*, 2–13. [CrossRef]
26. INCOSE. Guide to the Systems Engineering Body of Knowledge (SEBoK), International Council on Systems Engineering (INCOSE). 2021. Available online: https://www.sebokwiki.org/wiki/Systems_Engineering_Overview (accessed on 9 June 2021).
27. Haberfellner, R.; Nagel, P.; Becker, M.; Büchel, A.; von Massow, H. *Systems Engineering*; Springer Nature Switzerland AG: Cham, Switzerland, 2019; ISBN 978-3-030-13431-0.
28. Pistikopoulos, E.N.; Barbosa-Povoa, A.; Lee, J.H.; Misener, R.; Mitsos, A.; Reklaitis, G.V.; Venkatasubramanian, V.; You, F.; Gani, R. Process Systems Engineering–The Generation Next? *Comput. Chem. Eng.* **2021**, *147*, 107252. Available online: https://www.sciencedirect.com/science/article/pii/S0098135421000302 (accessed on 9 June 2021). [CrossRef]
29. Sturdivant, R.L.; Chong, E.K. Systems engineering of a terabit elliptic orbit satellite and phased array ground station for IoT connectivity and consumer internet access. *IEEE Access* **2016**, *4*, 9941–9957. Available online: https://ieeexplore.ieee.org/abstract/document/7565492 (accessed on 9 June 2021). [CrossRef]
30. Madni, A.M.; Madni, C.C.; Lucero, S.D. Leveraging digital twin technology in model-based systems engineering. *Systems* **2019**, *7*, 7. Available online: https://www.mdpi.com/2079-8954/7/1/7 (accessed on 9 June 2021). [CrossRef]
31. Zhang, X.; Lovati, M.; Vigna, I.; Widén, J.; Han, M.; Gal, C.; Feng, T. A review of urban energy systems at building cluster level incorporating renewable-energy-source (RES) envelope solutions. *Appl. Energy* **2018**, *230*, 1034–1056. Available online: https://www.sciencedirect.com/science/article/pii/S0306261918313503 (accessed on 9 June 2021). [CrossRef]
32. Frank, A.G.; Dalenogare, L.S.; Ayala, N.F. Industry 4.0 technologies: Implementation patterns in manufacturing companies. *Int. J. Prod. Econ.* **2019**, *210*, 15–26. Available online: https://www.sciencedirect.com/science/article/abs/pii/S0925527319300040 (accessed on 9 June 2021). [CrossRef]
33. Culot, G.; Orzes, G.; Sartor, M.; Nassimbeni, G. The future of manufacturing: A Delphi-based scenario analysis on Industry 4.0. *Technol. Forecast. Soc. Chang.* **2020**, *157*, 120092. Available online: https://www.sciencedirect.com/science/article/abs/pii/S0040162520309185 (accessed on 9 June 2021). [CrossRef]
34. Zheng, T.; Ardolino, M.; Bacchetti, A.; Perona, M. The applications of Industry 4.0 technologies in manufacturing context: A systematic literature review. *Int. J. Prod. Res.* **2021**, *59*, 1922–1954. [CrossRef]
35. Borowski, P.F. Innovative processes in managing an enterprise from the energy and food sector in the era of industry 4.0. *Processes* **2021**, *9*, 381. Available online: https://www.mdpi.com/2227-9717/9/2/381 (accessed on 9 June 2021). [CrossRef]
36. Qiu, W.; Li, B. Empirical Study of HRM Effectiveness of Research Teams Based on 5G Network and Internet of Things. *Microprocess. Microsyst.* **2021**, *80*, 103577. Available online: https://www.sciencedirect.com/science/article/abs/pii/S0141933120307274 (accessed on 9 June 2021). [CrossRef]
37. Liao, Y.; Deschamps, F.; Loures, E.D.F.R.; Ramos, L.F.P. Past, present and future of Industry 4.0-a systematic literature review and research agenda proposal. *Int. J. Prod. Res.* **2017**, *55*, 3609–3629. [CrossRef]

38. Ghobakhloo, M. Industry 4.0, digitization, and opportunities for sustainability. *J. Clean. Prod.* **2020**, *252*, 119869. Available online: https://www.sciencedirect.com/science/article/abs/pii/S0959652619347390 (accessed on 9 June 2021). [CrossRef]
39. Furstenau, L.B.; Sott, M.K.; Kipper, L.M.; Machado, E.L.; Lopez-Robles, J.R.; Dohan, M.S.; Cobo, M.J.; Zahid, A.; Abbasi, Q.H.; Imran, M.A. Link between sustainability and industry 4.0: Trends, challenges and new perspectives. *IEEE Access* **2020**, *8*, 140079–140096. Available online: https://ieeexplore.ieee.org/abstract/document/9151934 (accessed on 9 June 2021). [CrossRef]
40. Luthra, S.; Kumar, A.; Zavadskas, E.K.; Mangla, S.K.; Garza-Reyes, J.A. Industry 4.0 as an enabler of sustainability diffusion in supply chain: An analysis of influential strength of drivers in an emerging economy. *Int. J. Prod. Res.* **2020**, *58*, 1505–1521. [CrossRef]
41. García-Muiña, F.E.; Medina-Salgado, M.S.; Ferrari, A.M.; Cucchi, M. Sustainability transition in industry 4.0 and smart manufacturing with the triple-layered business model canvas. *Sustainability* **2020**, *12*, 2364. Available online: https://www.mdpi.com/2071-1050/12/6/2364 (accessed on 9 June 2021). [CrossRef]
42. Mansell, P.; Philbin, S.P.; Konstantinou, E. Delivering UN Sustainable Development Goals' Impact on Infrastructure Projects: An Empirical Study of Senior Executives in the UK Construction Sector. *Sustainability* **2020**, *12*, 7998. [CrossRef]
43. Alyamani, R.; Long, S. The Application of Fuzzy Analytic Hierarchy Process in Sustainable Project Selection. *Sustainability* **2020**, *12*, 8314. [CrossRef]
44. Telukdarie, A.; Munsamy, M.; Mohlala, P. Analysis of the Impact of COVID-19 on the Food and Beverages Manufacturing Sector. *Sustainability* **2020**, *12*, 9331. [CrossRef]
45. Abatecola, G.; Surace, A. Discussing the Use of Complexity Theory in Engineering Management: Implications for Sustainability. *Sustainability* **2020**, *12*, 10629. [CrossRef]
46. Scott, C.; Ahsan, M.; Albarbar, A. Machine Learning Based Vehicle to Grid Strategy for Improving the Energy Performance of Public Buildings. *Sustainability* **2021**, *13*, 4003. [CrossRef]
47. Andrews, D.; Newton, E.J.; Adibi, N.; Chenadec, J.; Bienge, K. A Circular Economy for the Data Centre Industry: Using Design Methods to Address the Challenge of Whole System Sustainability in a Unique Industrial Sector. *Sustainability* **2021**, *13*, 6319. [CrossRef]

Article

Delivering UN Sustainable Development Goals' Impact on Infrastructure Projects: An Empirical Study of Senior Executives in the UK Construction Sector

Paul Mansell [1,2,*], Simon P. Philbin [1] and Efrosyni Konstantinou [2]

1 Nathu Puri Institute for Engineering & Enterprise, London South Bank University, London SE1 0AA, UK; philbins@lsbu.ac.uk
2 Bartlett School of Construction and Project Management, University College London, London WC1E 6BT, UK; efrosyni.konstantinou@ucl.ac.uk
* Correspondence: paulrmansell@gmail.com

Received: 7 August 2020; Accepted: 19 September 2020; Published: 27 September 2020

Abstract: Achievement of the United Nations' 2030 Sustainable Development Goals (SDG) is of paramount importance for both business and society. Across the construction sector, despite evidence that suggests 88% of those surveyed want to measure the SDG impact at both the business and project levels, there continues to be major challenge in achieving this objective. This paper shares the results of a qualitative research study of 40 interviews with executives from the United Kingdom (UK) construction industry. It was supported by a text-based content analysis to strengthen the findings. The results indicate that SDG measurement practices are embraced in principle but are problematic in practice and that rarely does action match rhetoric. While the research was completed in the UK, the findings have broader applicability to other countries since most construction firms have extensive global business footprints. Researchers can use the findings to extend the current understanding of measuring outcomes and impact at project level, and, for practitioners, the study provides insights into the contextual preconditions necessary to achieve the intended outcomes of adopting a mechanism for the measurement of SDGs. The international relevance of this research is inherently linked to the global nature of the SDGs and therefore the results could be used outside of UK.

Keywords: sustainability; project success; business–society; business models; Sustainable Development Goals (SDGs); sustainable development; infrastructure project

---

1. Introduction

The establishment of any society rests on the development of a number of integrated areas, including industrial, social and economic systems [1], while consuming vast amounts of resources that often negatively impact the environment that they depend on. Many of these developments can be aligned directly or indirectly with projects delivered across infrastructure categories within the construction industry. Indeed, recent projections in the sector indicate that an estimated USD $94 trillion [2] of investment in infrastructure projects is required globally by 2040. This represents a massive opportunity to stimulate economic prosperity, reduce poverty and raise standards in health, education and gender equality. However, there are also risks that the infrastructure project investment is squandered ineffectively or, worse, damages the environment and society that the economic development is dependent on. According to Morris [3], given the critical role that the project management community play, there is an urgent need for further research to ascertain more effective strategies to ensure balanced sustainable development to counter the threats of climate change and other global goals. Such global goals have been codified through the UN's Sustainable Development Goals (SDGs).

In 2015, the international community responded to the sustainable development challenge with their report *Transforming Our World: The 2030 agenda for sustainable development* [4]. The SDGs are the United Nations' blueprint, with 193 nations signatories, to address the global challenges, such as poverty, inequality, climate change, environmental degradation, prosperity, peace and justice [5]. The concept of sustainable development acquired its most cohesive definition in the United Nations' 1987 Brundtland Commission report, which described it as "development that meets the needs of the present without compromising the ability of future generations to meet their own needs" [6]. Using the "triple bottom line" [7–9], Ochieng, Price and Moore [10] took the definition further by placing it in the context of global construction projects and describing it as the balance of economic, social and environmental aspects. In their book [10], they identify a number of systemic issues, "hard and soft" in nature, that provide new challenges for global construction projects in relation to sustainable development.

There is a continuing need to understand how sustainability and, specifically, performance against the SDGs can be measured for infrastructure projects from the construction sector. Recent research [11–14] has indicated that linking infrastructure project success to SDG targets is problematic. Also, evidence has indicated that, at the project manager level, whilst the appetite for action is very strong, especially by millennials [14], there do not appear to be the tools, methods, leadership or business-society-environment framework to carry out meaningful measurement of SDG success at project level. This represents a knowledge gap that results in weaker investment decisions since SDG lessons are not being learned from project delivery successes and failures. According to this development, the objectives of this research study are to (1) identify the challenges and opportunities of SDG measurement for construction-industry projects and (2) understand how the SDGs and more generally sustainability are viewed in the context of the corporate strategy of construction sector enterprises.

The following section provides the literature review, which includes a brief overview of the concept of SDGs and their linkage to sustainability theory as well as a review of challenges and opportunities for measuring SDGs in the construction industry. The literature review is used to synthesise a series of theory-driven propositions. This is followed by the methods section and subsequently the findings and discussion sections. The final section concludes the paper with evaluation of the propositions and proposals for critical success factors that might inform the development of a prototype model for the measurement of SDGs. This section also recommends areas for further studies. In summary, the objective of this research is to explore the contextual issues that affect the linking of global goals to local delivery on infrastructure projects. The specific research question is "How do senior leaders in the construction sector rate and use global UN SDGs for infrastructure investment decisions at the local level?" Several propositions are derived from the literature review and they explore the research question further. Importantly, whilst the research is based on interviews with senior executives of UK firms, they were representative of firms that mostly had a global or regional footprint (57%), had staff levels mostly in the range of 10–25,000 (62%) and were mostly at, or above, senior executive level (defined as having "director" in their role title), including nearly a third at CEO or board level who reflected individuals who could represent their firm's views. It is therefore considered that the value of this research has international relevance because of the inherent global nature of SDGs and the global footprint of the organisations interviewed.

2. Literature Review

In order to address the aforementioned objectives, five key themes were identified that impact the context of the use of SDGs in infrastructure projects.

2.1. Sustainable Development Goals

The most significant global response to the planetary boundary challenge was in 2015, when all governments ratified the UN's 17 Sustainable Development Goals [4] (as shown in Figure 1), to be achieved by 2030 (with 169 targets and 244 indicators also agreed in 2017). This represented a major step change in the implementation of the sustainability agenda and effective responses to the planetary boundary challenge. Although the SDGs build on the earlier Millennium Development Goals (MDGs) [4] by focusing on similar issues, the SDGs differ from the MDGs because they are for all countries in the world to implement, developed and developing alike [15]. Also, unlike the MDGs, the SDGs are focused on monitoring, evaluation and accountability across society, not just at national level, which is why it is critical that the link is made from the "bottom to top", meaning from the delivery of project-level impacts that can be assessed against the national and global targets and indicators. The research presented later shows this cannot currently be achieved, and the evidence [16,17] illustrates that the golden thread from project measurement to national/global level is missing. There is a gap between theory and practice [11,13].

Figure 1. The United Nations 17 Sustainable Development Goals [4] (full details can be accessed at https://sustainabledevelopment.un.org/). (Usage of graphic agreed by UN).

In order to understand why there is a perceived gap, it is helpful to analyse the structural build of the SDG performance framework. In this regard, the SDG delivery targets are understandably ambitious and needed a reporting framework that would drive meaningful and verifiable progress towards the 2030 targets. In 2017, the UN's Inter-agency Expert Group on Targets and Indicators for Sustainable Development designed a mechanism that linked goals, targets and indicators across the geographic and governance boundaries at national, regional and global levels [18]. Within this framework, shown in Figure 2, the Expert Group designed thematic areas that could also be used at the subnational level, but, because the targets and indicators were originally designed to be used at the global, regional and national levels, they had reduced applicability at organisational or project levels. Considering the aforementioned literature, it is possible to synthesise the first proposition related to the Sustainable Development Goals as follows. Based on this discussion, the first proposition was developed, shown below.

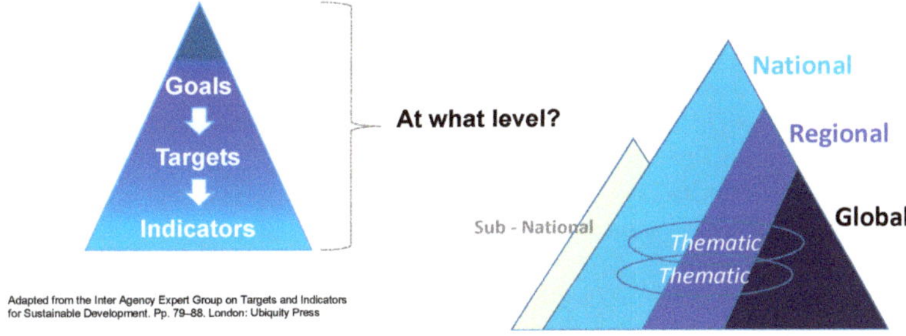

Figure 2. The Sustainable Development Goal (SDG) target and indicator framework developed by the UN subcommittee [18].

Proposition 1 (P1). *There is currently a gap in the knowledge base in regard to understanding how to measure SDG performance at the project level.*

2.2. SDGs in the Context of Infrastructure Projects in the Construction Sector

Most of society's developments in recent times can be connected to infrastructure projects [19,20] and the UN recognise that the development of infrastructure represents a massive opportunity to stimulate economic prosperity, reduce poverty and raise standards in health, education and gender equality [21]. It is apparent that ameliorating many of the risks associated with grand challenges, such as climate change, can only be achieved through investment in appropriate and resilient infrastructure and engineering [22].

A growing area of research has been in the comparison of construction projects' impacts on sustainable development from different angles. For example, Shen et al. [23] highlighted the role of projects to impact across the triple bottom line of people, profit and planet [7–9]. In this regard, construction projects are acknowledged as making an impact on the economic and social development of nations. Increasingly, recognition is given that these dual aims of economic development and social development can be achieved in harmony and, indeed, provide competitive advantage for firms [24,25]. Other studies have delved deeper into the changing nature of how project sustainability has changed within the construction industry. For example, Edum-Fotwe and Price [26] highlighted the issues that affect the assessment of social factors of construction projects, which this article suggests can be combined with the environmental and economic requirements of projects.

Defining infrastructure project success is central to understanding how to link global-national level SDGs with local infrastructure projects because it allows stakeholders to align their expectations against shorter-term outputs as well as the longer-term outcomes and SDG impacts. More recent research into project success definition [27] has consistently identified benefits and outcomes as being a critical determinant for the assessment of project success. Considering the aforementioned literature, it is possible to synthesise the second proposition related to SDGs in the context of infrastructure projects in the construction sector as follows.

Proposition 2 (P2). *The definition of infrastructure project success should be viewed from a systemic perspective, where there is a broader consideration of the overall performance of the project.*

2.3. Challenges and Opportunities for Measuring SDGs in the Construction Industry

As discussed above, there is evidence of an increasing interest, and in some cases demand, for promoting SDG measurement in the construction industry [19,20], with one report [14] that surveyed 325 engineers having a 95% demand from practitioners, who said that this was "very important" to them, with only 30% stating that they had adequate tools, processes and systems to measure them at project level. The survey [14] indicated four primary shortfalls for measuring SDGs on infrastructure projects, namely, leadership (1), tools and methods (2), engineers' business skills in measuring SDG impact (3) and how project success is too narrowly defined as outputs (such as time, cost and scope) and not outcomes (longer-term local impacts and stakeholder value) (4). This highlights that there are several challenges that impede the practical measurement of SDGs on projects, which need to be fully acknowledged.

Whilst there is still a limited body of research on the limitations of SDG measurement, there is much that can be learned from the measurement of sustainability on projects, and this is transferable to the SDG research. For example, Arif et al. [28] identified that there is often limited sustainability knowledge, especially amongst senior leaders, and this results in weaker understanding and impact assessments of related themes, such as poverty, environmental issues, supply chain adherence to sustainability best practice, cultural evaluation, technological deficiencies and limitations of research in depth and breadth, all of which have a negative influence on the valuation of sustainability, both as an investment lens and a delivery approach. A further barrier to the use of SDGs, which potentially mirrors sustainable construction, is what some authors have suggested is a lack of capacity and capability to implement effective and efficient sustainability [29]. Considering the aforementioned literature, it is possible to synthesise the third proposition related to SDGs in the context of the challenges and opportunities for measuring SDGs in the construction industry as follows.

Proposition 3 (P3). *Although there is knowledge of the importance of sustainability on infrastructure projects, there is a lack of awareness of how to measure the performance of infrastructure against the SDGs.*

2.4. The Concept of the Triple Bottom Line in Relation to SDGs

A contribution to the growing literature on the measurement of infrastructure projects on sustainability is provided by Ding and Shen [30], who focus on the balance needed between benefits to society whilst protecting the environment and still achieving the economic benefits envisaged in the project business case. The linkage across the three areas in the construction industry is further defined by Kibert [25], who suggests that the interrelationship between a project's outputs and the society that is impacted is a central component of defining the sustainability success of an infrastructure project. This introduces the concept that project success definition needs to consider success against the triple bottom line (TBL) [7–9] of social, environmental (or ecological) and economic (or financial) effects, otherwise noted as the "three pillars" concept of "people, profit and the planet" [7–9]. However, the overemphasis on the last of the TBL criteria, namely finance, brings us to the root of the problem of measuring projects' SDG impact [16,17].

This is because the crux of the project reporting problem lies with the dominance of accounting tools, which have been the preeminent business method of reporting business success for over 500 years since Luca Paccioli first published his papers on double entry bookkeeping [31]. It has largely remained unchanged until the past 10 years. As evidence of this widening to cover the three pillars of TBL, there has been a proliferation of mechanisms and economic models to track different elements of TBL, for example, environmental, social and governance (ESG) [7], which introduces these three core areas into the business investments decisions that measure the ethical and sustainability impacts of a company. The contention of this current research study is that the proliferation of project success measurement theories, tools and concepts, which are mostly finance-driven, causes confusion and often leads to suboptimal action [32] and that a TBL perspective needs to be integrated from the start of any business case development (see later section on business cases). Considering the aforementioned literature, it is possible to synthesise the fourth proposition related to SDGs in the context of the concept of the triple bottom line in relation to SDGs as follows.

Proposition 4 (P4). *Measurement of SDG performance should accommodate the perspective of the triple bottom line (i.e., social, environmental and economic performance).*

2.5. The Concept of Theory of Change in Relation to SDGs

There is a wide use of the Theory of Change (ToC) across many academic disciplines, including environmental and organisational psychology, but it has also increasingly been connected to sociology and political science. ToC emerged from the field of programme theory and programme evaluation in the mid-1990s as a new way of analysing the theories motivating programmes and initiatives working for social and political change. It is focused not just on generating knowledge about whether a programme is effective but also on explaining what methods it uses to be effective. The original work in the 1980s has been developed further by the work of notable methodologists, such as Huey Chen's work on theory-driven evaluations [33,34], Peter Rossi's systematic approach to theory-driven evaluation in social sciences [35], Michael Patton's focus on integrating the theory with practice [36,37] and Carol Weiss' seminal work that takes a stakeholder-centric perspective [38–42] to find more effective ways of evaluating complex community programmes.

Weiss suggests [38] that complex community programmes had not sufficiently aligned local stakeholders on the change process and what the outcomes will be. She noted that the logic chains are particularly weak in the midsection of the causal chain, without which the longer-term goals are weakened. Weiss uses the term "Theory of Change" to describe the causal links across the inputs–outputs–outcomes pathway. She also focused attention on what users could claim in terms of impacts, separating claims of "attribution" from a wider, less direct, "contribution". Based on her work [38–42], ToC has been applied extensively across international development, public health and human rights and has since become a central theory that underpins the approach to project benefits management [43–46].

The literature review has highlighted the potential benefits and tensions of linking global goals to local delivery on infrastructure projects. As a result of these findings, the derived research question is the following: how do senior leaders in the construction sector rate and use global UN SDGs for infrastructure investment decisions at the local level? The sub-questions that flow from this are as follows.

- What issues influence the successful use of an SDG measurement mechanism to achieve the desired outcomes? (This represents the context).
- What mechanism (for measuring SDG impacts) is in place to achieve the outcomes? (This represents the mechanism).
- What are the expected outcomes of successfully using the SDG measurement mechanism? (This represents the outcome).

Considering the aforementioned literature, it is possible to synthesise the fifth proposition related to the concept of the Theory of Change in relation to SDGs as follows.

Proposition 5 (P5). *Measurement of SDG performance should include a full project lifecycle perspective and take account of longer-term project outcomes and wider impacts.*

3. Methodology

The broader research design involved a three-way data collection approach (Figure 3). At its core, the research design built on the triangulation of qualitative and quantitative datasets, which is well recognised as a method for informing theory-led research development [47,48]. In what Creswell [47] describes as a sequential explanatory design, the literature review informs the survey questions and analysis that has informed the structure and approach of the interviews discussed in this article. In this way, Merriam and Grenier [49] suggest that "the interviews help the researcher understand the responses to the survey [14] as well as provide additional insights into the phenomenon of interest".

As shown in Figure 3, the development of a prototype SDG measurement model was to be based on the triangulation of learning from the literature review, the survey of 325 engineers and the subsequent interviews of 40 senior executives. Only the interview stage is shared in this paper. A primary advantage of the semi-structured interview method is that it allows an adaptive–responsive approach to ensure the best improvisation to delve deeper into relative areas of importance based on the participants' responses [50] and it also allows for participants' verbal expressions to be captured [51].

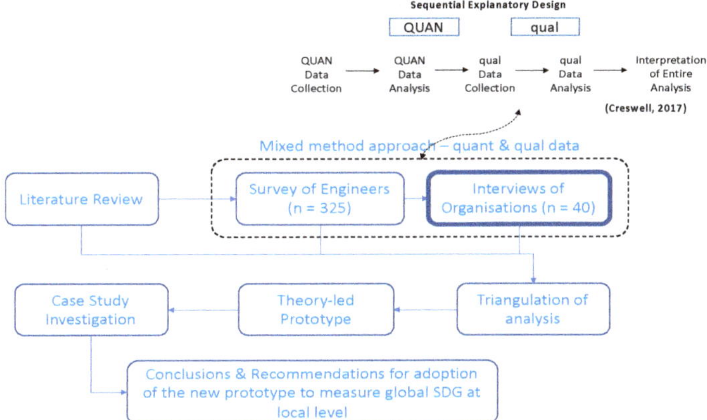

Figure 3. The research design of mixed-method sequential explanatory design, adapted from Creswell [47].

3.1. Using the Realist Evaluation Methodology to Structure the Survey

The study adopted the critical realism perspective of ideological philosophers, such as Bhaskar [52], to inform the choice of the realist evaluation approach primarily because of its practical utility and widespread use in social science research [53]. Taking Bhaskar's view [52], critical realism assumes that certain events exist and people then apply different perspectives and meaning to their interpretation of the truth.

3.2. Interview Question Design

The semi-structured interviews were designed to measure attitudes in relation to the research question and its subsidiary three sub-questions (shown in Figure 4). The sub-questions focused on three areas: the perceived value and importance of measuring SDGs (i.e., the outcomes), their current approach and capability (i.e., the mechanism) and their identification of the challenges and opportunities (i.e., the context), such as skills, tools, processes, structures and methods [54]. NVivo© was chosen as the web-enabled data collection tool.

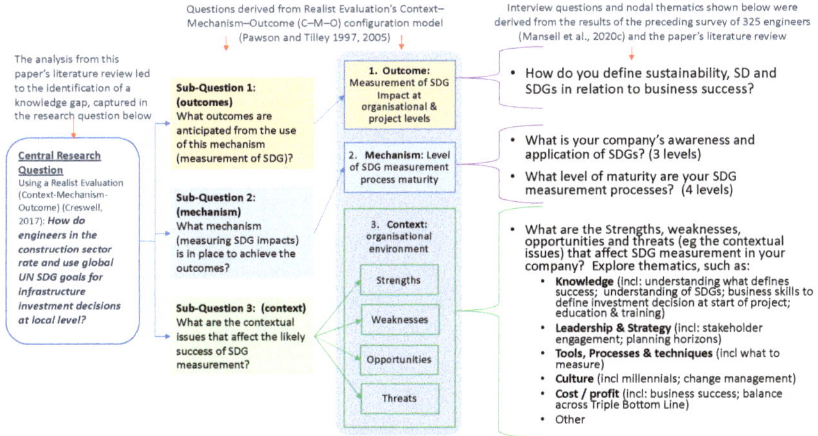

Figure 4. The nodal evaluation framework for the sequential explanatory design from which the semi-structured interview questions were derived.

3.3. Derivation of the Questions

The questions that are shown in Figure 4 have been derived from a variety of sources, both inductively and deductively. The central research question was informed by the literature review, which highlighted a knowledge gap. The importance of understanding why the gap existed and how to close the gap had also been identified by a previous survey of 325 engineers [14], in which 88% of responses affirmed that stakeholders wanted to increase their ability to measure SDGs on projects. This was strengthened by a response rate of only 34% stating that they had a "fit-for-purpose" mechanism to measure the SDG impacts [14]. The sub-questions 1–3 shown in Figure 4 were derived from the adoption of the realist evaluation's context–mechanism–outcome (C–M–O) configuration [55,56], which is widely used across clinical research (Pawson et al. 2005) and increasingly also across the social sciences [53]. Pawson and Tilley specifically recommend the C–M–O strategy so that "programme theories can be tested for the purposes of refining them" [55,57]. In this regard, the investigation is not about what works but asks instead "what works for whom in what circumstances and in what respects, how?" [55,57]. The third level of questions for the interviews (shown in the right column of Figure 4) combines the Pawson and Tilley C–M–O framework [55,57] with the survey results [14]. For example, the four contextual questions that were derived from the SWOT analysis were all topical responses from the surveys that engineers had identified as either "blockers" or opportunities [12].

3.4. Access

The interviews aimed to gain access to 40 CEOs or heads of sustainability. Given the GDPR issues around accessing the names of the senior executives of global companies, the research partnered with the Institution of Civil Engineers (ICE). The ICE vetted the research scope and agreed to provide the personal data on the basis of the work aligning with GDPR legalities. The lead researcher contacted a total of 85 organisations at the level of CEO and heads of sustainability, of which 40 agreed to be interviewed.

3.5. Sample Size

Sampling was achieved purposefully by partnering with UK's leading construction standards body, the Institution of Civil Engineers (ICE), to identify and select leaders in construction companies who had demonstrated a willingness to be involved in innovative knowledge development. All the interviewees had significant knowledge of the infrastructure sector but often did not have the detailed knowledge of their sustainability, SDG and CSR approaches. For this reason, the sample included 30% that were heads of sustainability, who had the requisite knowledge.

4. Data Analysis

4.1. Descriptive Statistics

The 40 interviews were conducted between July and September 2018, although two of the interviews had to be cancelled and the participants submitted their answers in writing. The descriptive statistical data are shown in Table 1.

The interviewees were representative of firms that mostly had a global or regional footprint (57%), as shown in Figure 5, had staff levels mostly from 1–25,000 (62%), and they were mostly at, or above, senior executive level (defined as having "director" in their role title), including nearly a third at CEO or board level who reflected individuals who could represent their firm's views.

Table 1. Profiles of participants.

Participant ID	Role in Company	Size of Company (Number of Employees)	Geography of Business	Length of Interview
1	Board	Other	Other	45
2	Senior executive	10–25k	Global	55
3	Head of sustainability	25–50k	Global	61
4	CEO	1–5k	Regional	42
5	CEO	1–5k	Regional	53
6	Senior executive	1–5k	Regional	53
7	CEO	>50k	Global	40
8	CEO	1–5k	National	42
9	Head of sustainability	1–5k	National	36
10	Senior government or UN policy director	1–5k	National	52
11	Senior executive	1–5k	National	36
12	CEO	5–10k	National	35
13	Senior executive	<1k	National	42
14	CEO	<1k	National	52
15	Head of sustainability	5–10k	Global	56
16	Board	5–10k	Global	56
17	Senior executive	>50k	Global	21
18	Senior government or UN policy director	Other	Other	36
19	Head of sustainability	10–25k	Global	75
20	Head of sustainability	10–25k	Global	55
21	Board	5–10k	Regional	45
22	Head of sustainability	1–5k	Regional	45
23	Head of sustainability	10–25k	Global	45
24	Senior executive	1–5k	Global	39
25	Senior executive	10–25k	National	43
26	Senior government or UN policy director	<1k	Global	38
27	Senior government or UN policy director	other	National	47
28	Senior executive	10–25k	Global	36
29	Head of sustainability	10–25k	National	46
30	Senior government or UN policy director	other	Other	65
31	Senior executive	<1k	National	59
32	Senior executive	<1k	National	59
33	Head of sustainability	10–25k	Global	43
34	Head of sustainability	10–25k	National	44
35	Board	10–25k	National	44
36	Senior executive	other	Global	65
37	Board	5–10k	Global	57
38	Head of sustainability	5–10k	Global	57
39	CEO	other	National	Written
40	CEO	other	Global	Written
			Total	1820
			Average	48

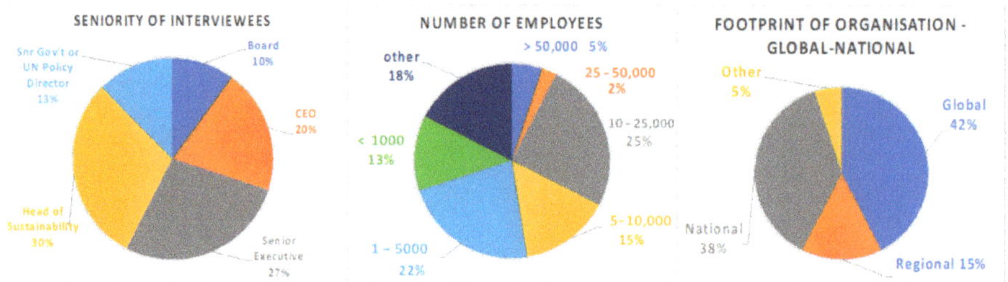

Figure 5. The attributes and values of the 40 interviewees.

4.2. Development of the Twin-Track Analysis Protocols, Balancing Qualitative with Quantitative Data Collection

As discussed in paragraph 3.3, the preferred approach was aligned to Frels and Onwuegbuzie [56], who had proposed that even within a specific method choice, such as interviews that are qualitative-dominant, it is appropriate to collect quantitative data during the qualitative interview process. The practical application of the "qualitative-dominant crossover" is shown below in Figure 6, which illustrates a twin track analysis method, which complemented the use of quantitative and qualitative data collection.

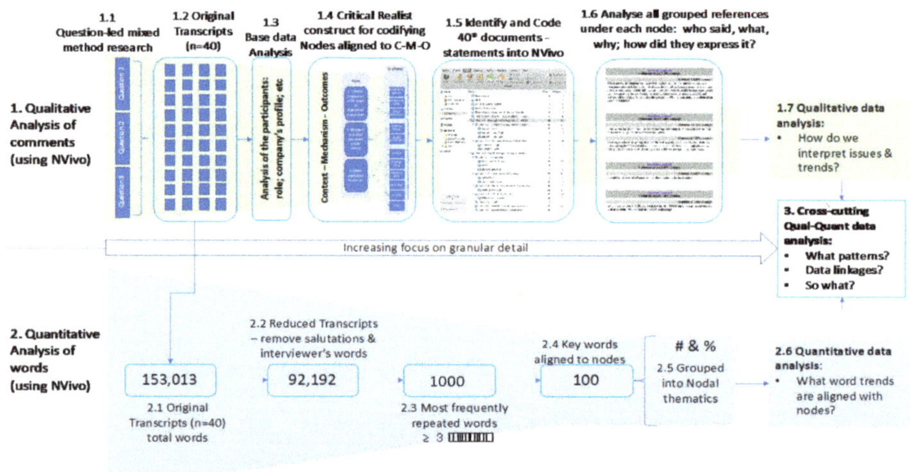

Figure 6. The twin-track analysis protocols approach: qualitative and quantitative.

4.3. Interview Analysis Process

All interviews were conducted in person and lasted an average of 48 min (min = 36; max = 75 min). With the participants' agreement, interviews were recorded using a digital recorder supplemented with hand-written notes. Later, the transcriptions, using the Trint© software tool, were uploaded onto NVivo© and were then compared and coded using the qualitative data analysis software.

The data were analysed at two levels, firstly analysed using textual analysis and then "made sense of" by using themes and pattern interpretation. Based on the nodal structure described earlier and using the parent-child branching technique (Figure 4), this provided an efficient and effective mechanism to capture and link themes but did not in itself provide any analysis. The nodal coding was aligned to the three research sub-questions, based on the realist evaluation C–M–O thematics [56,57], and each transcript was coded at three levels: first, second and third level coding (Figure 7). The frequency of participants' statements that were selected for coding, and also the relative frequency of nodal use. These groupings of statements under each node were then analysed for similarities and aligned with emerging themes.

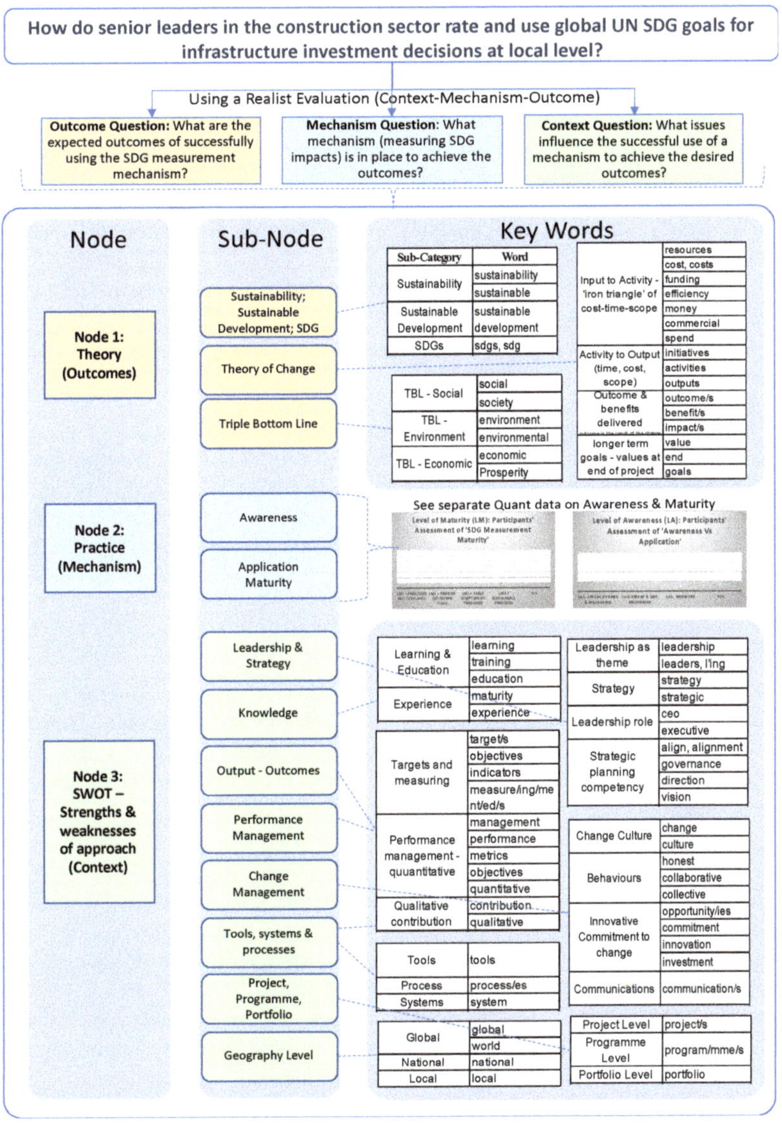

Figure 7. The nodal framework used for identification of key words aligned to context–mechanism–outcome (C–M–O) [55,56].

In addition to the primary analysis approach discussed above, the researchers complemented this with text mining analysis. This is a commonly used methodology for social scientists [58] because it enables the researchers to manage and quantify huge amounts of data in a very short time.

4.4. Verification

The verification was completed after the interpretation of the data analysis. This involved presenting the findings in workshops to leading practitioners within the confines of the standards body knowledge team at the Institution of Civil Engineers.

5. Results

The results and discussion are structured in three sections that relate to the three sub-questions, as shown in Figure 4, that stem from the primary research question: how do senior leaders in the construction sector rate and use global UN SDG goals for infrastructure investment decisions at the local level? The subsections are therefore as follows.

- Thematic area 1: outcome. What are the expected outcomes of successfully using the SDG measurement mechanism?
- Thematic area 2: mechanism. What design criteria enable the mechanism (for measuring SDG impacts) to achieve the outcomes?
- Thematic area 3: context. What issues influence the successful use of an SDG measurement mechanism to achieve the desired outcomes?

Using the twin track analysis approach (Figure 6), which includes both the qualitative and quantitative data, results are derived from the combined findings. All participants were asked for their views on the strengths, weaknesses, opportunities and threats (SWOT) of the employment of the mechanism. Given the semi-structured interview approach, their responses did not take a standard route and the interviewer used the funnelling technique [58] to increase subject specificity where depth of answer was required.

5.1. Thematic Area 1: Outcome. What Are the Expected Outcomes of Successfully Using the SDG Measurement Mechanism?

The "Outcome" section is the first of three thematic areas that focuses on the broader organisational ambitions of sustainability, sustainable development and SDGs. The results are collated under the following headings: the challenge/problem, the opportunities and the imperative for change. This thematic collected the second highest (out of 23 nodes and sub-nodes) number of references (n = 81) in NVivo for business views on the expected outcomes.

5.1.1. The Challenge/Problem

The essence of the problem was articulated by participant 10: "The weaknesses of the impact measures relate to some of the quantification of it in that there is no standard way of doing it and therefore quantifying impact is very difficult .... The leadership is not fully bought into it. It could be you have not got good sufficient tools for learning and education behind it. There is a lack of consistency in the data of how you measure it and the people measure it in different ways and people will have different perspectives of what good looks like". These views are similar to those of participant 26, who also noted the level of complexity, especially when positioned in a global context with the inherent cultural variations, which is potentially why so many participants only claimed to measure the SDGs at a high level: "This is so complex and it is so different if we are doing things in different countries with different organisations across different environments".

5.1.2. Overarching Opportunity

There were many participants that identified opportunities for improvement, and these are mostly captured under section three on "Contexts". The ambition, noted by many, was summed up by participant 26, who was from an international organisation and who gave this insight into his global organisation's aim: "In three years' time we would like to be in a position to have enough information based on evidence and frameworks in place so that we can have better conversations earlier on with clients about what the potential benefits are for the project and why we should be doing projects possibly in a different way than given to us by donors and others." He continued by anticipating the broader causal impacts of having this mechanism in place: "So, if we understand the linkages contribution projects can have across several SDGs, and how that impact could be measured, then we can have better conversations to understand where people should be investing their money and how, and what other aspects to bring into our project to ensure long-term sustainability". This places emphasis on using the SDGs to make better investment decisions, which becomes one of the critical success factors of the employment of this mechanism.

5.1.3. Imperative for Change and Commitment to Measure SDGs

Participant 13 explained the key part that SDGs contribute to the company's approach to the broader sustainability agenda: "The SDGs and our impacts on them are of huge relevance to our industry. We are already fully committed to measuring our impact across the triage of economic, social and environmental sustainability themes. Our leadership is fully committed to owning delivery success against these targets, which we jointly assess with our tier 1 contractors. It is now considered core business to ensure the right levels of scrutiny and governance to manage sustainable development performance. In future, this will include measurement against SDG targets but, for now, we need to find a practical method for doing this well." The final comment in the extract highlights the difficulty of moving from "knowing to doing".

Many commented on the link between SDG measurement and their company's values. For example, participant 5: "because our purpose is far more than simply generating revenues for shareholders ... for us, it is about influencing those solutions to provide the right long-term infrastructure for society. So, we provide jobs and the right training and we provide the infrastructure we need to connect life together; everything we do depends on it—to try to capture the way we go about doing that in more modern ways for future societies". Although many were better able to relate progress stories with their sustainability measurement, there were others, such as participant 28: "the whole world has decided how it can be rapidly made better, so the 169 SDG targets are a compass for humanity".

The theme of creating shared value [59,60] was commented on by a number of participants (2, 5, 8, 10, 11, 13 and 19), one of whom, a CEO, commented, "Since becoming Responsible Business of the Year, we have been working hard to show others how sustainability makes good business sense." This quote emphasises that the notion of creating shared value (CSV) [59,60], whilst not always using the specific language of CSV, is a growing reason to engage with SDGs and sustainability more generally.

The global context and the relationship of the global SDG goals to businesses was a common theme, as indicated by participant 24, head of infrastructure for his company, who said "in a world where populations are increasing, cities are expanding and the effect on our environment is more apparent than ever before, the need for infrastructure that is affordable, sustainable and effective is vital. Engineers have a pivotal role to play in designing infrastructure that is not only effective but does not harm the environment in which we live".

The first major finding derived from this analysis is as follows.

Findings #1: to achieve the outcomes of measuring SDG impacts at subnational level, business priorities can be aligned across economic, environment and society ambitions, and it can make good business sense to do this.

5.2. Thematic Area 2: Mechanism. What Mechanism (for Measuring SDG Impacts) Is in Place to Achieve the Outcomes?

The second area of discussion was for the participant to self-assess their company's "awareness and application" and also, if they were applying SDGs, what the level of process maturity of their SDG measurement was. The data in Figure 8 show the feedback from the participants when they were asked to score themselves against a Likert-style scale, as shown in the first row in columns c and d.

a	b	c	d	e	f	g	h
Participant ID	**Role in Company**	**Awareness Vs Application** (1= aware & doing; 2= aware & not doing; 3= unaware & not doing)	**Your Company's Level of Maturity in SDG Measurement** (0= process not developed; 1= definition developing; 2= early processes in place; 3= using sustainable processess)	**Participant ID**	**Role in Company**	**Awareness Vs Application**	**Level of Maturity**
1	Board	n/a	n/a	21	Board	2	1
2	Senior Executive	1	2	22	Head of Sustainability	2	1
3	Head of Sustainability	1	3	23	Head of Sustainability	2	0
4	CEO	2	1	24	Senior Executive	2	0
5	CEO	1	2	25	Senior Executive	1	1
6	Senior Executive	1	2	26	Senior Government or UN Policy Director	2	1
7	CEO	3	0	27	Senior Government or UN Policy Director	1	n/a
8	CEO	2	0	28	Senior Executive	1	2
9	Head of Sustainability	1	2	29	Head of Sustainability	2	1
10	Senior Government or UN Policy Director	2	1	30	Senior Government or UN Policy Director	2	0
11	Senior Executive	1	2	31	Senior Executive	n/a	n/a
12	CEO	1	2	32	Senior Executive	n/a	n/a
13	Senior Executive	2	0	33	Head of Sustainability	2	1
14	CEO	1	1	34	Head of Sustainability	2	0
15	Head of Sustainability	2	1	35	Board	2	2
16	Board	1	1	36	Senior Executive	2	n/a
17	Senior Executive	1	2	37	Head of Sustainability	2	1
18	Senior Government or UN Policy Director	n/a	n/a	38	Head of Sustainability	2	1
19	Head of Sustainability	1	3	39	CEO	n/a	n/a
20	Head of Sustainability	1	3	40	CEO	n/a	n/a

Figure 8. Results of the self-assessed level of awareness-application and process maturity (colour representation shown in columns c and d in titles row).

5.2.1. Company's "Awareness and Application" of SDG Measurement in Construction Projects

As part of the interviews, all participants were asked to describe their awareness of sustainability, sustainable development and SDGs. They were then asked to describe their current level of SDG measurement maturity. The data on these are shown in Figure 9.

At the lower end of the spectrum (level 3 = unaware and not doing it), participant 37 admitted that, regarding "the United Nations Sustainable Development Goals, I had never heard of them—a request for an interview came through and [name withheld] only heard of them through a bid we were working on that included an SDG question. The SDGs have no current place in our business". As this was a board member, this was surprising because it was expected that senior staff would have some level of SDG knowledge.

In the middle range, which was "aware and not doing", representing 47% of the participants, participant 4's answer was typical: "Awareness is that we are doing some discrete things but not in any depth". The reasons for this varied, but a common theme was that there was not a requirement from governments or clients, as participant 21 shares: "We do not have a demand from our clients or from our communities that we work to measure against the SDGs. Like many in our industry, these are not common terms that we use ... we do not have as much benefit from embedding them as much as a large global company that perhaps needs to demonstrate SDG impact more visibly. A lot of the things we do implicitly encompass the SDGs, but we are not explicitly measuring against them".

In the higher range, which was "aware and measuring", represented by 38% of the participants, there were some examples of significant progress, such as that shared by participant 7: "Every single project in the organisation will feed into SDG number 11—'sustainable cities and communities'—and every project in the organisation will address at least 4–5 of the SDGs".

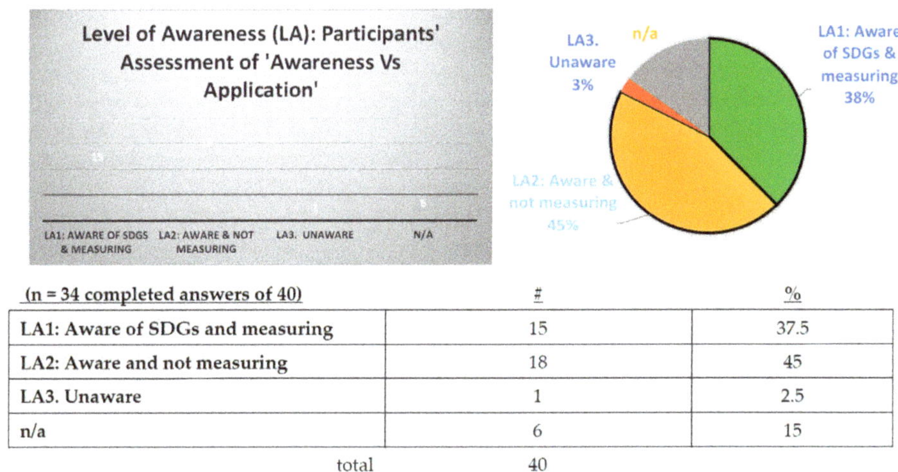

(n = 34 completed answers of 40)

	#	%
LA1: Aware of SDGs and measuring	15	37.5
LA2: Aware and not measuring	18	45
LA3. Unaware	1	2.5
n/a	6	15
total	40	

Figure 9. Graphical representation of results from the self-assessed level of "awareness-application".

5.2.2. Company's Level of SDG Measurement Process Maturity

As part of the interviews, the second quantitative question all participants were asked was to describe their current level of SDG measurement maturity. The data on these are shown in Figure 10. The banding levels for this question were: 0 = no SDG processes, 1 = currently defining processes, 2 = early processes in place and 3 = sustainable SDG processes. Overall, the quantitative data showed that nearly half (49%) were at level 0 or level 1, which meant that no effective processes were in operational use. Only 23% stated that they were at level 2, the early adoption stage of processes, with a very small group (8%) stating that they had repeatable processes in place.

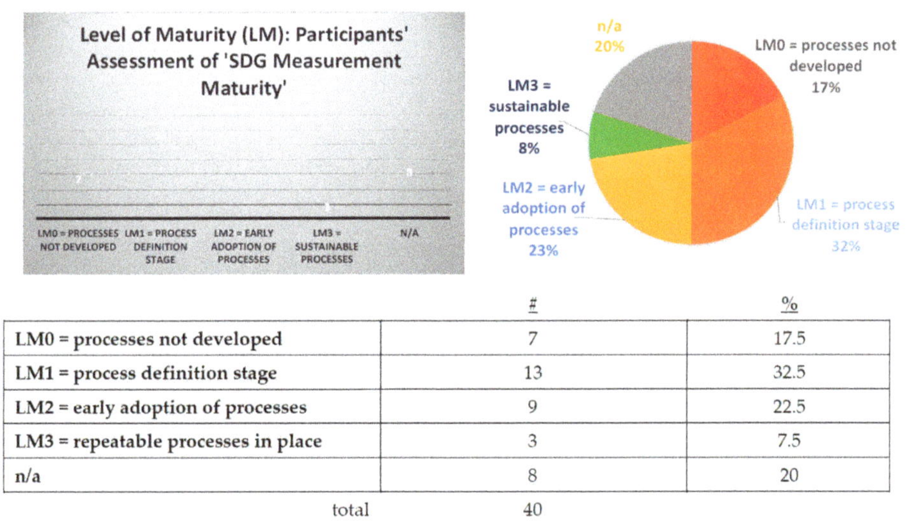

	#	%
LM0 = processes not developed	7	17.5
LM1 = process definition stage	13	32.5
LM2 = early adoption of processes	9	22.5
LM3 = repeatable processes in place	3	7.5
n/a	8	20
total	40	

Figure 10. Graphical representation of results of the self-assessed level of "SDG measurement maturity" (colour representation shown in Figure 8 in column h in the titles row).

One of the best, participant 13, stated: "We are at Level 3, we have managed processes, metrics and quality management", which was similar to participant 23: "we have some consistent ways we do things that are aligned to SDGs, but we do not look at every SDG and answer how they contribute to the goals. But we do cover a lot of the issues at project level."

In reality, many of the participants only conducted measurement at a high level, such as participant 34: "In the past we have done a review to see how our strategy fits with the SDGs. We found that the SDGs were impacted by our work, some more than others, in terms of the goals and targets; they are not particularly relevant to the work that we do so our priorities have been elsewhere and therefore our resources have been focused elsewhere". About a third of the participants said that they could, at a high level, link their SDG priorities to the formal sustainability reporting that they did on the Global Reporting Index (GRI), such as participant 26, who stated: "Well, we are all aware and starting to do it. We started using the Global Reporting Index framework on sustainability three years ago and we started reporting on our corporate results yearly on that but, at the project level, we have been a bit slower pushing up to that". Amongst the lowest performers was participant 9, who stated: "in terms of SDG reporting processes we are close to 1. Our maturity is still low, although our sustainability reporting is much higher. We have not yet made it integrated to SDGs and have not yet generated a report against them. That is what we are talking about now and what we want to achieve".

The second major finding derived from this analysis is as follows.

Findings #2: only a small percentage of companies have a repeatable process as an operational "mechanism" for measuring SDG impacts at company and project levels. Most have an aspiration to do so but believe that the government and their clients need to require its implementation.

5.3. Thematic Area 3: Context. What Issues Influence the Successful Use of an SDG Measurement Mechanism to Achieve the Desired Outcomes?

The analysis of the contextual issues that affect companies' ability to measure SDG impacts successfully were captured using a strength, weakness, opportunity and threat (SWOT) approach. The eight themes are shown in the nodal framework in Figure 7 and include: leadership and strategy; knowledge; outputs-to-outcomes; tools, processes and systems; change management; performance management; project-to-portfolio levels; and geographic issues. These were all derived from the preceding survey of 325 engineers, as shown in Figure 3. The qualitative analysis shared below is complemented by using the twin-track approach described in Figure 6, which includes the text-analysis software-enabled word-count data. The approach was to identify key words and relate their frequency of use to the qualitative findings to assist the understanding of the emerging issues. For example, in this first context thematic, "leadership and strategy", as shown in Table 2, the key words associated with this thematic are: leadership (and its derivatives, such as leader), strategy, CEO/executive and align/governance/direction/vision, which are all words associated with leadership capabilities and actions.

Table 2. Text analysis (NVivo) on key words' frequency: context of leadership.

Category	C-M-O	Sub-Category	Word	Count	f1 Word %	f2 Sub-Cat %	f3 Cat %
Leadership	Context	Leadership as theme	leadership	83	0.16%	0.29%	0.80%
			leaders	30	0.06%		
			leading	20	0.04%		
			leads	15	0.03%		
		Strategy	strategy	75	0.14%	0.18%	
			strategic	23	0.04%		
		Leadership role	CEO	26	0.05%	0.12%	
			executive	37	0.07%		
		Strategic planning competency	align	25	0.05%	0.21%	
			governance	25	0.05%		
			alignment	18	0.03%		
			direction	18	0.03%		
			vision	28	0.05%		

5.3.1. Leadership and Strategy

For the leadership and strategy node, there were high levels of relevant statements coded (n = 63) from the 40 participants (using the NVivo software), reflecting the importance of this thematic. In terms of key word usage, this thematic was the fifth most frequently used (n = 584) across the 40 interviews, which equates to once every 120 words. Within this category, the frequency of use of "align", "governance", "direction" and "vision" were noted since these words are all associated with leadership capabilities.

The most impactful statements collected were the frequent references to a "greater value" beyond profit. This sentiment sits well with creating shared value and the triple bottom line discussed earlier. This viewpoint was personified by participant 11: "a key part of leadership is doing the right thing because it is the right thing to do, not because of a box-ticking exercise". The same participant also focused on the difficulty of making the change stick: "It is 50% belief and 50% belligerence when you start something like this; that is, holding yourself and others to account. That is what I mean by belligerence. In other words, 'seeing it through' and what we wrote down as a mantra: 'Don't you understand'". In his view, as a senior executive, he stressed the important role of his CEO and board: "Leadership is the most important critical success factor, both internally and externally, to align and galvanise our employees, our communities and the supply chain. It was about getting us all to be more collaborative in finding novel, innovative ways of delivering sustainable solutions. It is about the leaders capturing the hearts and minds of the stakeholders to champion changed behaviours to achieve big, bold strategic outcomes."

In terms of strategy, one organisation noted the importance of the "ends, ways, means" logic similar to the Theory of Change concept [38–42]. Participant 9 stated: "you must start with the end in mind, even if you have not got a detailed route map to deliver at every stage of the journey. Part of the mantra is to set big audacious goals and then adopt an attitude of 'I have started so I will finish' and, by the way, you never actually finish, because the end goal is moving, it is like you achieve one peak but realise it is a false horizon, and so you continue your climb to the next summit". The value of having clarity of the strategic ends is noted, albeit with a caution that the identification of targets for tracking performance must not become a "box-ticking" exercise that distorts clarity of outcomes. Participant 11 stated: "if you actually begin with the end in mind of the outcome you are seeking and how you wire your DNA to achieve that, you are far more likely to achieve those outcomes and, in so doing, the boxes get ticked. But if you predicate your thinking with thoughts about just filling the boxes, you have constrained yourself".

Finding #3: strong leadership plays a significant part in inculcating SDG measurement as an ambition and core value into an organisation.

Finding #4: the more advanced businesses in SDG measurement noted the need to have a clearly defined strategy that can guide the prioritisation of SDG goals using the "ends, ways, means" model. This requires clarity of the "ends" prior to defining project success (in-project and post-project).

5.3.2. Knowledge

For the "knowledge" node there was a relatively smaller incidence (n = 19) of relevant statements coded from the 40 participants (using the NVivo software). In terms of key word usage, as shown in Table 3, this thematic was also one of the least frequently used, with "learning", "education" and "experience" being used only 140 times across the 40 interviews, which equates to once every 400 words.

The qualitative analysis identified a strong preference for using education and training to improve their staff's SDG impact skills and business skills, especially in the wider definition of success, which is related to the later discussion of outputs-to-outcomes. An indication of the importance of this was provided by the CEO of one global engineering company, participant 7: "So, how do we galvanise our community, how do we tell our story better against the SDGs and how do we galvanise our community to be able to share best practices, and what does that mean for education and training?"

Table 3. Text analysis (NVivo) on key words' frequency: context of knowledge.

Category	C-M-O	Sub-Category	Word	Count	f1 Word %	f2 Sub-Cat %	f3 Cat %
Knowledge	Context	Learning & Education	learning	30	0.06%	0.15%	0.26%
			training	22	0.04%		
			education	29	0.05%		
		Experience	maturity	30	0.06%	0.11%	
			experience	29	0.05%		
				140	0.26%		

Skills covered a number of areas, including the skills to be able to define success definitions, business skills to be able to build performance frameworks and sustainability/SDG skills that helped understand the SDG framework and how they relate at sub global-national levels and at organisational and project levels. Participant 3 stressed its import: "I think the skills piece is the second most important area because we cannot expect our people to deliver on these KPIs if they do not know what they mean and if they do not know how to measure them and improve them, so investing in how to calculate social value and improve upon them and investing in training in social value RoI is very important; it gives us an opportunity to benchmark and improve on it".

Overall, participants seemed to accept that, despite the current supposed level of SDG measurement awareness, there is also a shortage of trained personnel to support the implementation of SDG measurement on their construction projects. The closing of this gap reflects the views of Reffat [61] on the insufficient number of human resources with the required skills to perform sustainable development on construction operations.

Finding #5: learning and education plays a critical role in increasing capability and, specifically, in understanding how to better share lessons on SDG measurement for the good of all.

5.3.3. Outputs-to-Outcomes

The "outputs-to-outcomes" node had the fifth highest incidence (n = 30) of relevant statements coded from the 40 participants (using the NVivo software). In terms of key word usage, this thematic was also one of the most frequently used (shown in the "Theory of Change" key word table, Table 4), with the first half of the causal chain (input, activities and outputs) being cited as frequently as the second half of the value chain (outcome to impacts). This was significantly less than the general reference to longer-term benefits that were synonymous with key words such as "value", "ends" and "goals", which were used only 339 times across the 40 interviews, equating to once every 175 words.

Within this subcode, most recognised the challenge of differentiating between outputs and outcomes. Too few knew how to do this well and, as a result, the wrong "targets indicators" were sometimes being used to measure success. Participant 8, a CEO of one of the UK's largest infrastructure programmes, said: "programme and project people are sometimes less aware of how we are doing strategically if you are not careful. So, they can often have a bias for cost and schedule focus and lose focus on other priorities we have set". Another way of expressing the inappropriate focus on outputs came from participant 3: "we know that, if we just design to code, we end up with projects that are great for today but absolutely do not meet the future that we are expecting".

Some organisations have fully embraced the strategic aim of better aligning with outcomes, such as participant 11: "So we thought long and hard not just about the goals that we created but about how they fitted with a set of outcomes in our region and what that would look like in terms of implementation. This was our way of meaningfully connecting the strategy with outcomes that our stakeholders recognised." The same person described the need to look at the end first to better understand ambitions: "you must start with the end in mind, even if you have not got a detailed route map to deliver at every stage of the journey". One of the most common reasons for the overemphasis on "outputs" was shared by participant 26: "So, the measurables are very weak in terms of linking the engineering and the infrastructure impacts to the higher programme. It is just about 'have you built the hospital' as an output".

Table 4. Text analysis (NVivo) on key words' frequency: mechanism/context of the Theory of Change.

<table>
<tr><th>Category</th><th>C-M-O</th><th>Sub-Category</th><th>Word</th><th>Count</th><th>f1 Word %</th><th>f2 Sub-Cat %</th><th>f3 Cat %</th></tr>
<tr><td rowspan="20">Theory of Change, (causal logic chain from inputs to impacts)</td><td rowspan="20">Mechanism</td><td rowspan="8">Input to Activity 'iron triangle' of cost-time-scope</td><td>resources</td><td>20</td><td>0.04%</td><td rowspan="8">0.33%</td><td rowspan="20">1.26%</td></tr>
<tr><td>cost</td><td>57</td><td>0.11%</td></tr>
<tr><td>costs</td><td>17</td><td>0.03%</td></tr>
<tr><td>funding</td><td>16</td><td>0.03%</td></tr>
<tr><td>efficiency</td><td>16</td><td>0.03%</td></tr>
<tr><td>money</td><td>18</td><td>0.03%</td></tr>
<tr><td>commercial</td><td>17</td><td>0.03%</td></tr>
<tr><td>spend</td><td>17</td><td>0.03%</td></tr>
<tr><td rowspan="3">Activity to Output (time, cost, scope)</td><td>initiatives</td><td>27</td><td>0.05%</td><td rowspan="3">0.12%</td></tr>
<tr><td>activities</td><td>15</td><td>0.03%</td></tr>
<tr><td>outputs</td><td>19</td><td>0.04%</td></tr>
<tr><td rowspan="6">Outcome and benefits as result of change derived from project's outputs.</td><td>outcomes</td><td>60</td><td>0.11%</td><td rowspan="6">0.17%</td></tr>
<tr><td>outcome</td><td>34</td><td>0.06%</td></tr>
<tr><td>benefits</td><td>23</td><td>0.04%</td></tr>
<tr><td>benefit</td><td>19</td><td>0.04%</td></tr>
<tr><td>impact</td><td>219</td><td>0.41%</td></tr>
<tr><td>impacts</td><td>19</td><td>0.04%</td></tr>
<tr><td rowspan="3">longer term goals—values at end of project</td><td>value</td><td>101</td><td>0.19%</td><td rowspan="3">0.64%</td></tr>
<tr><td>end</td><td>73</td><td>0.14%</td></tr>
<tr><td>goals</td><td>165</td><td>0.31%</td></tr>
</table>

Finding #6: the use of the log-frame and Theory of Change provides a means to link outputs to outcomes and better identify SDG impacts.

5.3.4. Tools, Processes and Systems

The "tools, processes and systems" node had one of the lowest incidences (n = 18) of relevant statements coded from the 40 participants (using the NVivo software). This suggests that senior executives and CEOs have less interest in, or place lower value on, specific tools or methodologies, which might indicate why this is an underinvested area. In terms of key word usage, this thematic was also one of the least frequently used, shown at Table 5, with "processes" being cited twice as frequently as "tools" and "systems". In total, they were used only 177 times across the 40 interviews, which equates to once every 300 words.

The survey [14] that preceded these interviews had identified a common reference to the lack of tools, systems and methodologies. This was not proven in the interviews, although a number of the heads of sustainability (3, 9, 15, 20 and 29) were more likely to mention this as a factor. On the ability of the sector to galvanise and align with a consistent approach, participant 18 highlighted that there were bigger issues to deal with prior to designing a tool: "I think it is essential. I have very little confidence in our ability to do it now. Even if you had a decent methodology now, I suspect very few people would use it and you probably have a number of competing methodologies, which is typical in this sector."

However, others, such as participant 20, said: "for me the tools and processes underpin the delivery because, without them, you cannot possibly know where you are or where you need to go". But a key element of the design of a tool was to get the balance right between being too complex and being at the other end of the scale—being too high level and therefore superficial—as noted by participant 10: "I think, in most cases, a consistent framework or reporting approach would be helpful; that gets the balance right between having something that is consistent but watered down to such a high level that it loses meaning, versus having too much detail that is too granular, loses the users in too much complexity and is difficult to fit with your business model and the way you report things into that".

Table 5. Text analysis (NVivo) on key words' frequency: tool, processes and systems.

Category	C-M-O	Sub-Category	Word	Count	f1 Word %	f2 Sub-Cat %	f3 Cat %
Tools, Systems; Processes	Mechanism	Tools	tools	32	0.06%	0.06%	0.34%
		Process	processes	26	0.05%	0.23%	
			process	93	0.18%		
		Systems	system	26	0.05%	0.05%	

Finding #7: the use of tools, systems and processes to measure SDGs is not a priority for CEOs and board members but it is for senior executives and heads of sustainability. These tools need to be simple enough to understand but robust enough to capture detailed evidence that leads to improved performance.

5.3.5. Change Management

The "change management" node had an average level of incidence (n = 27) of relevant statements coded from the 40 participants (using the NVivo software). In terms of key word usage, this thematic (shown in Table 6) tracked "change culture", "behaviours", "innovation" and "communications", all of which provided a large number of insights from participants. In total, they were used 410 times across the 40 interviews. However, the quantification of the data does little to indicate that this contextual issue was one of the best sources of insightful knowledge.

There was general recognition from the participants that the single most important area for ensuring SDG measurement success is having a successful change programme that ensures a practical approach is made to work for the "users", with the added value of what they are doing. The starting point for this approach was ensuring the right culture in the organisation, characterised by openness and honesty about the difficulties of measuring SDGs and also closing the gap between superficial statements of intent without having the evidence to back up what they say they do. For example, participant 15 stated: "[name of company removed] say that they measure against SDGs, but there is a gap between what they say they do and what they actually do".

Innovation was a frequently referenced benefit of getting the change culture right and, in doing so, having the means to address the SDG targets more effectively. For example, participant 11 noted the effect of building long-term supplier relationships that enabled more innovative solutions to be developed: "We wanted to establish meaningful change across the supply chain, and we recognised that, to do this, we had to develop long-term relationships; hence, we contracted on a five, plus five, plus five-year basis. This built longevity into our thinking and allowed true innovation to develop solutions to the bigger sustainable development issues across the environment, driving efficiency and effectiveness."

Communication was also a dominant theme of culture change. Participant 1 noted: "you do not communicate it once, you communicate it nearly every day through many, many different vehicles. You bring people in". Participant 24, a leader of a North American national civil engineers institution, highlighted the value of leaders who can tell stories that resonated with stakeholders: "people with success stories become your spokespeople and they start to influence others, saying 'hey, you know this works for us' rather than just trying to sell the methodology. It is more, you know, encouraging peers, e.g., peer-to-peer". The main focus for this stakeholder engagement for participant 11 was: "Our starting point is understanding what is important to our clients, who want to see us make improvements, and where our staff and employees want to make a difference".

An unexpected but often-quoted issue was on the context of gender influence on SDG measurement. Eight participants (1, 5, 10, 17, 21, 24, 31 and 37) made specific reference to gender impact: "the younger generation really do want to change the world. Interestingly, particularly the female part of that [company name removed] has more than 50% of its membership as female and I pondered why that should be, and I think it is because it appeals to the values of certainly the younger, but actually to the female, side of our institution, who really want to make a difference to the world that they live in. Probably, they are more driven by that than they are by financial reward".

There were nearly half the participants that promoted the positive effects of harnessing the power of the millennial generation to promote change and thereby help champion the uptake of SDG measurement, which was shared by participant 1: "So, if we can find a way of linking into the power of the younger generation". This attitude was further explored by participant 10, who noted the obvious fact that millennials are tomorrow's leaders: "I think millennials have a role here as new project leaders where often they are the people who are most energised".

Table 6. Text analysis (NVivo) on key words' frequency: context of change management.

Category	C-M-O	Sub-Category	Word	Count	f1 Word %	f2 Sub-Cat %	f3 Cat %
Change Management	Context	Change Culture	change	129	0.24%	0.28%	0.78%
			culture	23	0.04%		
		Behaviours	honest	22	0.04%	0.10%	
			collaborative	14	0.03%		
			collective	14	0.03%		
		Innovative Commitment to change	opportunity	54	0.10%	0.34%	
			opportunities	20	0.04%		
			commitment	42	0.08%		
			innovation	32	0.06%		
			investment	32	0.06%		
		Communication	Communication /s	28	0.06%	0.06%	

Finding #8: change management. One of the largest positive impacts for SDG measurement is about engaging, communicating and energising the delivery teams. This involves the internal teams and suppliers. The millennials have a key role to help build and sustain this change momentum.

5.3.6. Performance Management

The "performance management" node had the highest level of incidence (n = 82) of relevant statements coded from the 40 participants (using the NVivo software). In terms of key word usage, this thematic, shown in Table 7, tracked "targets", "measuring", "performance management", "quantitative", "metrics", "qualitative" and "contribution". In total, they were used 1003 times across the 40 interviews, which equates to once every 50 words and represents the most referenced thematic.

The highest frequency of coding on NVivo was using the node for "what to measure", reflecting the importance of this thematic. There were many references to what is measured, and the general theme was that the selection of targets becomes critical in a business environment that is already awash with data collection. Many asked whether they should collect quantitative data or qualitative and also asked what the balance between too little data collection and too much is. Almost all participants accepted that this was an extremely difficult area to resolve and that there were no easy answers. For example, participant 34 stated: "I think we are quite confused. It sounds like we are much more advanced than we are in the way we monitor, report and evaluate. Most of our work is about getting the basics right and ensuring we are complying with legal requirements—getting stuff done. We know we need to do more work on understanding sustainability outcomes and how we can develop detailed KPIs that feed into that for measuring our impact. We do not have outcome frameworks in place yet".

There was a consistent recognition amongst those that had more advanced levels of SDG measurement process maturity (participants 3, 19 and 20) that you had to start by selecting a manageable number of goals (from 17) and targets (from 169). This was explained by participant 31, who said: "It is an enormous challenge. I think, out of those 232, the fact that you found 20 that can be measured is actually pretty good if I think about the magnitude of the problem". Amongst the nine participants that were at the "early processes in place" stage, most were trying to establish hard metrics that could be quantified, such as participant 15: "We want hard targets to test our performance. Generally, as a business, qualitative is not very compelling. When we set up our strategy, we did some serious baselining to get some better referenced data."

One of the key problems, mentioned earlier, is the level of complexity in measuring 169 SDG targets. It was frequently explained that this was too complicated for the construction sector, as stated by participant 2: "But the indicators are far too detailed and big and sometimes not applicable as well. Therefore, it is better to work at a higher level for the projects. I have more interest in the goals and not the indicators".

The emphasis on quantifiable targets was countered by participant 25: "telling the story of the success against the sustainable development goals, as an example; a lot of the time, it cannot be quantified very easily and therefore telling the story around an outcome perhaps provides more impact and value than just putting a meaningless quantitative score against something". This viewpoint was backed by participant 2: "In the beginning, I wanted quantification to have numbers that I can use to understand the measurement data. This created a big pushback because engineers tend to want perfect solutions. The assessment was causing some culture issues, so the qualitative aspects have been preserved but not the quantitative. So, we still look for the holy grail but, at this stage, we are going to produce stories. In future we would like more quantitative that can be assessed at corporate level."

Table 7. Text analysis (NVivo) on key words' frequency: performance management.

Category	C-M-O	Sub-Category	Word	Count	f1 Word %	f2 Sub-Cat %	f3 Cat %
Performance Management	Mechanism	Targets and measuring	target	31	0.06%	1.36%	1.90%
			targets	208	0.39%		
			objectives	26	0.05%		
			indicators	76	0.14%		
			measure	142	0.27%		
			measuring	72	0.14%		
			measuring	72	0.14%		
			measurement	54	0.10%		
			measured	17	0.03%		
			measures	21	0.04%		
		Performance management - quantitative	management	83	0.16%	0.43%	
			performance	54	0.10%		
			metrics	46	0.09%		
			objectives	26	0.05%		
			quantitative	18	0.03%		
		Qualitative contribution	contribution	40	0.08%	0.11%	
			qualitative	17	0.03%		

Finding #9: select a few targets relevant to the construction organisation or project. Keep it simple and build knowledge progressively.

5.3.7. Project-to-Portfolio Levels

The "project-to-portfolio" node had the eighth highest level of incidence (n = 21) of relevant statements coded from the 40 participants (using the NVivo software). In terms of key word usage, this thematic (shown in Table 8) tracked "projects", "programmes" and "portfolios". In total, they were used 677 times across the 40 interviews, which equates to once every 80 words. There was wide recognition that the approach needed to be adapted but linked across the project, programme and portfolio levels, as noted by participant 27: "I think there is no 'one size fits all'. So, I think it will vary from programme to programme and be dependent on the country as well".

Special interest and importance were aligned with the node on "starting projects". The preceding survey [14] had not highlighted the importance of "starting projects well". This node was added during the interviews stage because it was often referred to as the need to use the SDG lens at the "key investment decision point", as noted by participant 26: "based on evidence frameworks, you can frame your project in a much better way to make sure the impact you get is maximized." The emphasis of getting stakeholder alignment was also mentioned by participant 19: "They want to demonstrate that their projects contribute to sustainability development goals and develop tools that make sure projects embed sustainability development at the outset, e.g., at their project inception phase".

There were some, such as participant 9, the head of sustainability for a utility company, who suggested that the SDG measurement had more relevance at the larger scale of programmes and at the organisational strategic level, represented by the portfolio office: "Thus we do it more at programme and portfolio level and less at project level. So, we have a mapping process at the portfolio level and align across project and programme SDG targets".

Table 8. Text analysis (NVivo) on key words' frequency: projects-to-portfolios.

Category	C-M-O	Sub-Category	Word	Count	f1 Word %	f2 Sub-Cat %	f3 Cat %
Project; Programme; Portfolio	Context	Project Level	project	278	0.52%	0.87%	1.27%
			projects	185	0.35%		
		Programme Level	program	65	0.12%	0.35%	
			program	65	0.12%		
			programme	31	0.06%		
			programs	27	0.05%		
		Portfolio Level	portfolio	26	0.05%	0.05%	

Finding #10: there was evidence that SDGs can be measured at all three levels: projects, programmes and portfolios. There was special value in using the SDG lens at the start of the project to help align stakeholders around the longer-term outcomes and impacts.

6. Discussion

This section builds on the 10 core findings and culminates with generalisations across the three sub-questions that guided the design of this research into SDG measurements. The three sub-questions, as shown in Figure 4, stem from the primary research question: how do senior leaders in the construction sector rate and use global UN SDG goals for infrastructure investment decisions at the local level? The empirical research study, including aforementioned qualitative findings and supporting quantitative data, also allows an evaluation of the theory-driven propositions to be undertaken, which is provided according to the following areas of outcome, mechanism and context.

6.1. Outcome Discussion: What Are the Expected Outcomes of Successfully Using the SDG Measurement Mechanism?

The results showed that participants have the appetite and resolve to employ SDG measurement at business and project levels (Finding #2) in order to achieve outcomes that benefit people, the planet and profit. At the same time, they were frustrated by their inability to do so for reasons discussed in the following sections. Most participants were optimistic that their organisation would achieve the broader outcomes by making SDG measurement more usable, consistent and verifiable across the construction sector, with increasing balance to their investment decisions across environment, economic and societal factors (Finding #1). There was almost unanimous conviction that the "ends" of achieving the desired "outcomes" was good for business (Finding #4).

Although the results emerged from a different thematic, some of the participants (2, 3, 17, 19, 20, 26 and 27) recognised the value of using Carol Weiss' seminal work [38–42] that uses the logframe and Theory of Change approach to take a stakeholder-centric perspective to assist the definition of longer-term impacts and outcomes. They acknowledged that this helps rebalance from an overemphasis on output definition, which is typically used in project management and too often judges success in terms of delivering the infrastructure asset to time, cost and scope (Finding #6).

The findings from the research study allow evaluation of the propositions synthesised from the literature review as follows.

Proposition 2 was supported through inference from the analysis.

Proposition 5 was supported.

6.2. Mechanism Discussion: What Design Criteria Enable the Mechanism (for Measuring SDG Impacts) to Achieve the Outcomes?

The views were consistent (with the four exceptions mentioned in the preceding paragraph) in stating that this was an important area for the construction sector to get right but that there was no best practice established for how to deliver an effective mechanism. Therefore, despite the strong support for its adoption, the depth of knowledge on SDGs was mostly superficial, and only 8% of the organisations interviewed self-assessed their SDG measurement processes as repeatable (Finding #2), with only a further 23% having processes at an "early adoption stage". The majority had not yet defined the SDG measurement processes. Unsurprisingly, there were many, especially at board and CEO level (with notable exceptions, such as 5, 7, 8 and 12), who showed some confusion in their knowledge of SDGs, sustainability and sustainable development. This was reflected in having relatively consistent and well-informed views on specialist areas, such as carbon management, but this was less evident in the details of what the SDGs represented.

The low level of uptake of the SDG measurements at the project level was attributed to the following reasons. (a) The complexity of the SDG framework, with the scale of ambition understandable at a high level but made excessively complicated when examining the 17 goals, 169 targets and 232 indicators. (b) The lack of adoption of SDGs by clients did not mandate SDG measurement (Finding #2). There was therefore no incentive to dedicate finite resources to a complicated task that might not deliver any value; indeed, it might even identify their weaknesses, which only a few explicitly opined was a good way of learning and developing.

A further design criterion that emerged, to enable the mechanism for measuring SDG impacts to achieve the outcomes, was the ability to find a golden thread from enterprise portfolio level to project level (Finding #10). This was most clearly explained by the participants that were most developed in their SDG measurement processes (2, 3, 11 and 20) but also included others who were actively developing SDG processes (8, 9, 14, 19, 27, 28 and 36). Whilst there was confidence in their self-assessed ability to achieve the golden thread from project to portfolio level (Finding #10), this was mostly not substantiated by any evidence (except 2, 3 and 11).

The findings from the research study allow evaluation of the propositions synthesised from the literature review as follows.

Proposition 1 was supported.

Proposition 4 was supported through inference from the analysis.

6.3. Context Discussion: What Issues Influence the Successful Use of an SDG Measurement Mechanism to Achieve the Desired Outcomes?

As part of the discussions on strengths and weaknesses, the participants identified a number of contextual issues that affected the likely success of the mechanism achieving the desired outcomes. These "context" issues included leadership (Finding #3), outcome-output definition (Finding #4), knowledge (Finding #5) and change management (Finding #8) capabilities. There were more optimistic discussions than pessimistic ones about the ways they could improve the contextual issues identified. However, a few had little incentive for, or perceived little value in, adding what they considered a burdensome task onto the shoulders of busy project managers.

Given the seniority of the participants, it was not surprising that leadership and strategy was a dominant theme in discussions. This led to Finding #3, which states that strong leadership plays a significant part in inculcating SDG measurement as an ambition and core value into an organisation. This was most clearly stated by a senior executive (11): "Leadership is the most important critical success factor, both internally and externally, to align and galvanise our employees, our communities and the supply chain". Others (2, 10, 17, 19 and 29), none of whom were CEOs or board members, stated that the strategic nature of organisational change had to be driven from the top [62]. There was recognition that, in reality, this meant that leaders at all levels were needed as champions, which, for SDG measurement, needed to be aligned with success stories that would make sense to the target audience, expressed in their language and justifying "why" followed by explaining clearly "how".

Linking to the models developed by Kotter [62] on leading change, the eighth finding was related to the contextual issue of change management (Finding #8). One of the most significant ways to influence the take-up of SDG measurement across organisations is engaging, communicating and energising the delivery teams. Research has shown that this is critical to achieving the right organisational cultures [63].

The findings from the research study allow evaluation of the propositions synthesised from the literature review as follows.

Proposition 3 was supported.

The contextual issues identified above are a small insight into broadening our understanding of factors that influence construction companies' decisions on whether to use SDGs as a lens for defining success and, if so, how they might use them effectively. Other studies delve deeper into construction sustainability benefits [64] or, for example, the evaluation of modern methods of construction based on wood (as aligned to SDG 12 on responsible consumption and production) [65]. Equally important areas that are not addressed in the thematics discussed above relate to green financing; some authors [66] have provided insights into public–private partnerships as a mechanism for financing sustainable development. This highlights the breadth of relevant thematics and keeps the focus of this paper on just the restricted areas considered most important to the executives interviewed.

7. Conclusions and Future Work

This comprehensive research study has provided empirically grounded insights from the 40 senior leaders on their perceptions of how their organisations rated and used SDGs as a measurement lens. The 10 findings have provided a rich and deep insight into answering the question of how to measure SDG performance on infrastructure projects. The empirical research has also validated the theory-driven propositions that were synthesised from the literature. Furthermore, this research study identifies that, whilst SDG measurement practices on infrastructure projects are embraced in theory, they are problematic in practice: rarely does action match rhetoric.

Although the 40 interviews described in the study specifically identified a primary stakeholder group, the senior executives of construction firms, there were a number of other stakeholders included, viz. two senior government experts in the infrastructure sector, one financial advisor, one from the United Nations and three from standards bodies. Consequently, the study seeks to include the considerations of wider stakeholders involved in project decision-making. The research team have also consulted with the UK's Institution of Civil Engineers to ensure this broader perspective is adequately captured.

There is evidence that, although the study was completed in the UK, the results may be applicable to a wider international group because most of the firms have extensive global footprints. It is therefore considered that the inherent global nature of SDGs and the global footprint of the organisations interviewed results in the broader international value of this research. The specific benefit to researchers is that the findings extend knowledge on the theory of measuring outcomes and impact at project level, and, for practitioners, the study provides insights into the contextual preconditions necessary to achieve the intended outcomes of adopting a mechanism for the measurement of SDGs. In this way, the article offers learning that has significant implications for investment decisions, where being able to systematically identify SDG impacts, from the start, is helpful for achieving local impact against global targets, with broader benefit for people, profit and the planet. The broader SDG research programme that this paper is part of has worked closely with many international organisations, such as UNOPS, which also signifies that this is an area that has wide relevance and can be added to the growing literature across the world on how we are addressing the grand challenges of the SDGs.

One of the primary characteristics of this qualitative research is that the researcher "is the primary instrument for data collection and data analysis" [49]. However, there is a paradox that, despite this strength, it is also a potential weakness since, unlike a survey or scientific experiment, this allows the "human instrument" to adjust to evolving changes. For example, the lead researcher allowed the interview questions to evolve in a free-flowing discussion when he noted that a different line of enquiry might provide unexpected new insights. There is thus a need to apply some caution to the potential hazard of bringing the researchers' own bias [49], since "it is important to identify them [bias and subjectivity] and monitor them as to how they may be shaping the collection and interpretation of data" (p.13). Another limitation of this study was the research approach. Further research could be expanded to include case studies that test the relevant SDG mechanisms to assess whether the outcomes can be achieved.

In regard to future research, there was a lack of evidence given by participants on their ability to achieve the golden thread of SDG measurement from project to portfolio level (Finding #10) because, often, it was not available at any credible depth or backed up by verifiable evidence. It is therefore proposed that this is an area for further research to test whether aspirations to achieve this linkage are realistic. There is also the need for further research outside the UK since, while the findings from this study have broad global application due to the regional and global footprint of the participants' organisations, the complexities and challenges in some areas require further SDG measurement research.

Author Contributions: The research was developed by several authors as follows: Conceptualization, P.M.; methodology, P.M.; validation, P.M., S.P.P. and E.K.; formal analysis, P.M.; investigation, P.M.; data curation, P.M.; writing—original draft preparation, P.M.; writing—review and editing, P.M., S.P.P. and E.K.; visualization, P.M.; supervision, S.P.P. and E.K.; project administration, P.M. All authors have read and agreed to the published version of the manuscript.

Funding: This research was indirectly funded by the Nathu Puri Institute for Engineering and Enterprise, School of Engineering, London South Bank University, through the funding of the doctoral research support to Paul Mansell.

Acknowledgments: The authors would like to thank the Institution of Civil Engineers (ICE) for their advice and support through this phase of research.

Conflicts of Interest: The authors declare no conflict of interest. The funders had no role in the design of the study; in the collection, analyses, or interpretation of data; in the writing of the manuscript or in the decision to publish the results.

References

1. Heravi, G.; Fathi, M.; Faeghi, S. Evaluation of sustainability indicators of industrial buildings focused on petrochemical projects. *J. Clean. Prod.* **2015**, *109*, 92–107. [CrossRef]
2. Global Infrastructure Hub. Infrastructure Investment Need in the Compact with African Countries. 2019. Available online: https://outlook.gihub.org/?utm_source=GIHub+Homepage&utm_medium=Project+tile&utm_campaign=Outlook+GIHub+Tile (accessed on 6 June 2020).
3. Morris, P.W.G. *Climate Change and What the Project Management Profession Should be Doing about It*; Association for Project Management: Regent Park, UK, 2017; Available online: https://www.apm.org.uk/media/7496/climate-change-report.pdf (accessed on 11 May 2018).
4. United Nations. *Transforming Our World: The 2030 Agenda for Sustainable Development*; Resolution Adopted by the General Assembly; United Nations: New York, NY, USA, 2015.
5. United Nations. World Population Prospects. 2019. Available online: http://esa.un.org/unpd/wpp/Publications/Files/Key_Findings_WPP_2015.pdf (accessed on 6 June 2020).

6. Brundtland, G.H. *Our Common Future: Report of the World Commission on Environment and Development*; Oxford University Press: Oxford, UK, 1987.
7. Elkington, J. Towards the sustainable corporation: Win-win-win business strategies for sustainable development. *Calif. Manag. Rev.* **1994**, *36*, 90–100. [CrossRef]
8. Elkington, J. Enter the triple bottom line. In *The triple Bottom Line*; Routledge: London, UK, 2013; pp. 23–38.
9. Elkington, J. 25 Years Ago I Coined the Phrase "Triple Bottom Line." Here's Why It's Time to Rethink It. Available online: https://hbr.org/2018/06/25-years-ago-i-coined-the-phrase-triple-bottom-line-heres-why-im-giving-up-on-it (accessed on 6 June 2020).
10. Ochieng, E.G.; Price, A.D.F.; Moore, D. *Management of Global Construction Projects*; Palgrave Macmillan's Global Academic: Hampshire, UK, 2013.
11. Mansell, P.; Philbin, S.P.; Broyd, T.; Nicholson, I. Assessing the impact of infrastructure projects on global sustainable development goals. *Eng. Sustain.* **2020**, *173*, 196–212.
12. Mansell, P.; Philbin, S.P. Measuring Sustainable Development Goal Targets on Infrastructure Projects. *J. Mod. Project Manag.* **2020**, *8*.
13. Mansell, P.; Philbin, S.P.; Broyd, T. Development of a New Business Model to Measure Organizational and Project-Level SDG Impact—Case Study of a Water Utility Company. *Sustainability* **2020**, *12*, 6413. [CrossRef]
14. Mansell, P.; Philbin, S.P.; Konstantinou, E. Redefining the Use of Sustainable Development Goals at the Organisation and Project Levels—A Survey of Engineers. *Adm. Sci.* **2020**, *10*, 55. [CrossRef]
15. Sachs, J.; Woo, W.T.; Yoshino, N.; Taghizadeh-Hesary, F. Importance of green finance for achieving sustainable development goals and energy security. In *Handbook of Green Finance: Energy Security and Sustainable Development*; Springer: Singapore, 2019; pp. 3–12.
16. Martens, M.L.; Carvalho, M. The challenge of introducing sustainability into project management function: Multiple-case studies. *J. Clean. Prod.* **2016**, *117*, 29–40. [CrossRef]
17. Martens, M.L.; Carvalho, M. Sustainability and Success Variables in the Project Management Context: An Expert Panel. *Proj. Manag. J.* **2016**, *47*, 24–43. [CrossRef]
18. Inter-Agency and Expert Group on SDG Indicators (IAEG-SDGs). Resolution Adopted by the General Assembly on Work of the Statistical Commission Pertaining to the 2030 Agenda for Sustainable Development (A/RES/71/313). 2017. Available online: https://undocs.org/A/RES/71/313 (accessed on 2 April 2019).
19. Thacker, S.; Hall, J. *Engineering for Sustainable Development*; Infrastructure Transition Research Consortium (ITRC), University of Oxford: Oxford, UK, 2018.
20. Thacker, S.; Adshead, D.; Fay, M.; Hallegatte, S.; Harvey, M.; Meller, H.; O'Regan, N.; Rozenberg, J.; Watkins, G.; Hall, J.W. Infrastructure for sustainable development. *Nat. Sustain.* **2019**, *2*, 324–331. [CrossRef]
21. United Nations Office for Project Services (UNOPS). *Infrastructure: Underpinning Sustainable Development*; UNOPS: Copenhagen, Denmark, 2018; Available online: https://www.itrc.org.uk/wp-content/PDFs/ITRC-UNOPS-Infrastructure_Underpining_Sustainable%20Development.pdf (accessed on 19 March 2019).
22. Organisation for Economic Co-Operation and Development (OECD). G20/OECD Principles of Corporate Governance. Paris. 2015. Available online: https://www.oecd-ilibrary.org/governance/g20-oecd-principles-of-corporate-governance-2015_9789264236882-en (accessed on 20 March 2019).
23. Shen, L.; Tam, V.W.Y.; Tam, L.; Ji, Y. Project feasibility study: The key to successful implementation of sustainable and socially responsible construction management practice. *J. Clean. Prod.* **2010**, *18*, 254–259. [CrossRef]
24. Griggs, D.; Stafford-Smith, M.; Gaffney, O.; Rockström, J.; Öhman, M.C.; Shyamsundar, P.; Steffen, W.; Glaser, G.; Kanie, N.; Noble, I. Policy: Sustainable development goals for people and planet. *Nature* **2013**, *495*, 305. [CrossRef] [PubMed]
25. Kibert, C.J. *Sustainable Construction: Green Building Design and Delivery*, 3rd ed.; Wiley: Hoboken, NJ, USA, 2013.
26. Edum-Fotwe, F.T.; Price, A.D.F. A social ontology for appraising sustainability of construction projects and developments. *Int. J. Proj. Manag.* **2009**, *27*, 313–322. [CrossRef]
27. Thiry, M. Value Management. In *Wiley Guide to Managing Projects*; Morris, P., Pinto, J., Eds.; Wiley: Holbroken, NJ, USA, 2004.
28. Arif, M.; Egbu, C.; Haleem, A.; Kulonda, D.; Khalfan, M. State of green construction in India: Drivers and challenges. *J. Eng. Des. Technol.* **2009**, *7*, 223–234. [CrossRef]

29. Hakkinen, T.; Belloni, K. Barriers and drivers for sustainable building. *Build. Res. Inf.* **2011**, *39*, 239–255. [CrossRef]
30. Ding, G.K.; Shen, L. Assessing sustainability performance of built projects: A building process approach. *Int. J. Sustain. Dev.* **2010**, *13*, 267. [CrossRef]
31. Yamey, B.S. Scientific bookkeeping and the rise of capitalism. *Econ. Hist. Rev.* **1949**, *1*, 99–113. [CrossRef]
32. Silvius, A.J.; Schipper, R.P. Sustainability in project management: A literature review and impact analysis. *Soc. Bus.* **2014**, *4*, 63–96. [CrossRef]
33. Chen, H.T.; Rossi, P.H. Evaluating with sense: The theory-driven approach. *Eval. Rev.* **1983**, *7*, 283–302. [CrossRef]
34. Chen, H.T.; Chen, H.T. *Practical Program Evaluation: Assessing and Improving Planning, Implementation, and Effectiveness*; Sage: Thousand Oaks, CA, USA, 2005.
35. Rossi, P.H.; Lipsey, M.W.; Henry, G.T. *Evaluation: A Systematic Approach*; Sage Publications: Thousand Oaks, CA, USA, 2018.
36. Patton, M.Q. Reports on Topic Areas: The Evaluator's Responsibility for Utilization. *Am. J. Eval.* **1988**, *9*, 5–24. [CrossRef]
37. Patton, M.Q. A world larger than formative and summative. *Am. J. Eval.* **1996**, *17*, 131–144. [CrossRef]
38. Weiss, C.H. The stakeholder approach to evaluation: Origins and promise. *New Dir. Prog. Eval.* **1983**, *1983*, 3–14. [CrossRef]
39. Weiss, C.H. Nothing as practical as good theory: Exploring theory-based evaluation for comprehensive community initiatives for children and families. In *New Approaches to Evaluating Community Initiatives: Concepts, Methods, and Contexts*; The Aspen Institute: Washington, DC, USA, 1995; Volume 1, pp. 65–92.
40. Weiss, C.H. *Evaluation: Methods for Studying Programs and Policies*, 2nd ed.; Prentice Hall: Upper Saddle River, NJ, USA, 1998.
41. Weiss, C.H. Have we learned anything new about the use of evaluation? *Am. J. Eval.* **1998**, *19*, 21–33. [CrossRef]
42. Weiss, C.H. Theory-Based Evaluation: Theories of Change for Poverty Reduction Programs. *Eval. Poverty Reduct.* **2018**, 103–112. [CrossRef]
43. Atkinson, R. Project management: Cost, time and quality, two best guesses and a phenomenon, its time to accept other success criteria. *Int. J. Proj. Manag.* **1999**, *17*, 337–342. [CrossRef]
44. Crawford, L. Senior management perceptions of project management competence. *Int. J. Proj. Manag.* **2005**, *23*, 7–16. [CrossRef]
45. PMI Standards Committee, and Project Management Institute. *A Guide to the Project Management Body of Knowledge*; Project Management Institute: Newtown Square, PA, USA, 1996.
46. Pinkerton, W.J.; Pinkerton, W.J. *Project Management: Achieving Project Bottom-Line Success*; McGraw-Hill Education: New York, NY, USA, 2003.
47. Creswell, J.W.; Clark, V.L.P. *Designing and Conducting Mixed Methods Research*; Sage Publications: Thousand Oaks, CA, USA, 2017.
48. Easterby-Smith, M.; Thorpe, R.; Lowe, A. *Management Research: An Introduction*, 2nd ed.; Sage Publications: London, UK, 2002; p. 342.
49. Merriam, S.B.; Grenier, R.S. *Qualitative Research in Practice: Examples for Discussion and Analysis*; John Wiley & Sons: Hoboken, NJ, USA, 2019.
50. Rubin, D.B. Causal Inference Using Potential Outcomes. *J. Am. Stat. Assoc.* **2005**, *100*, 322–331. [CrossRef]
51. Berk, M.L.; Schur, C.L.; Cantor, J.C. Ability to Obtain Health Care: Recent Estimates from the Robert Wood Johnson Foundation National Access to Care Survey. *Health Aff.* **1995**, *14*, 139–146. [CrossRef]
52. Bhaskar, R. *A Realist Theory of Science*; Routledge: Oxford, UK, 2013.
53. Linsley, P.; Howard, D.; Owen, S. The construction of context-mechanisms-outcomes in realistic evaluation. *Nurse Res.* **2015**, *22*, 28–34. [CrossRef]
54. Astbury, B.; Leeuw, F.L. Unpacking Black Boxes: Mechanisms and Theory Building in Evaluation. *Am. J. Eval.* **2010**, *31*, 363–381. [CrossRef]
55. Pawson, R.; Tilley, N. Realistic evaluation bloodlines. *Am. J. Eval.* **2001**, *22*, 317–324. [CrossRef]
56. Frels, R.K.; Onwuegbuzie, A.J. Administering quantitative instruments with qualitative interviews: A mixed research approach. *J. Couns. Dev.* **2013**, *91*, 184–194. [CrossRef]
57. Pawson, R.; Tilley, N.; Tilley, N. *Realistic Evaluation*; Sage Publications: Thousand Oaks, CA, USA, 1997.

58. Kallio, H.; Pietilä, A.M.; Johnson, M.; Kangasniemi, M. Systematic methodological review: Developing a framework for a qualitative semi-structured interview guide. *J. Adv. Nurs.* **2016**, *72*, 2954–2965. [CrossRef] [PubMed]
59. Porter, M.E. *Creating and Sustaining Superior Performance. Competitive Advantage*; Free Press: New York, NY, USA, 1985; p. 167.
60. Porter, M.E.; Kramer, M.R. The Big Idea: Creating Shared Value, Rethinking Capitalism. *Harv. Bus. Rev.* **2011**, *89*, 62–77.
61. Reffat, R. Sustainable construction in developing countries. In Proceedings of the First Architectural International Conference, Cairo University, Cairo, Egypt, 24–26 February 2004.
62. Kotter, J.P. *Leading Change*; Harvard Business Press: Brighton, MA, USA, 2012.
63. Garavan, T.; McGuire, D. Human resource development and society: Human resource development's role in embedding corporate social responsibility, sustainability, and ethics in organizations. *Adv. Dev. Hum. Resour.* **2010**, *12*, 487–507. [CrossRef]
64. Švajlenka, J.; Kozlovská, M.; Pošiváková, T. Analysis of Selected Building Constructions Used in Industrial Construction in Terms of Sustainability Benefits. *Sustainability* **2018**, *10*, 4394. [CrossRef]
65. Švajlenka, J.; Kozlovská, M. Perception of User Criteria in the Context of Sustainability of Modern Methods of Construction Based on Wood. *Sustainability* **2018**, *10*, 116. [CrossRef]
66. Sergi, B.S.; Popkova, E.G.; Borzenko, K.V.; Przhedetskaya, N.V. Public–Private Partnerships as a Mechanism of Financing Sustainable Development. In *Financing Sustainable Development*; Palgrave Macmillan: Cham, Switzerland, 2019; pp. 313–339.

Article

The Application of Fuzzy Analytic Hierarchy Process in Sustainable Project Selection

Rakan Alyamani * and Suzanna Long

Department of Engineering Management & Systems Engineering, Missouri University of Science and Technology, Rolla, MO 65409, USA; longsuz@mst.edu

* Correspondence: rar3d@mst.edu

Received: 5 September 2020; Accepted: 5 October 2020; Published: 9 October 2020

Abstract: The project selection process is a crucial step in sustainable development. Effective sustainable development depends on the ability to select the appropriate sustainable project to implement to ensure that the desired goals are met. Some of the most common characteristics or criteria used in evaluating sustainable projects include novelty, uncertainty, skill and experience, technology information transfer, and project cost. Prioritizing these criteria based on relative importance helps project managers and decision makers identify elements that require additional attention, better allocate resources, as well as improve the selection process when evaluating different sustainable project alternatives. The aim of this research is to use the fuzzy analytic hierarchy process (FAHP) methodology in which fuzzy numbers are utilized to realistically represent human judgment to rank the different project criteria based on relative importance and impact on sustainable projects. The results from the FAHP show that the most important criterion to consider in sustainable project selection is project cost, followed by novelty and uncertainty as the second and third most important criteria, respectively. The two least important criteria out of the total of five examined in this research were the skill and experience and technology information transfer, respectively. These results will help project managers and decision makers identify selection criteria with higher weights of importance. Given that the selection criteria chosen for this research are not limited to the evaluation of a specific type of sustainable projects or a specific location, they can be used to evaluate different types of sustainable projects in different environments and locations.

Keywords: fuzzy analytic hierarchy process; project selection; sustainable projects; multi-criteria decision making

---

1. Introduction

The use of fossil fuels as a source of energy has been linked to a wide range of issues such as geographical dependency, limited resources, and low efficiency [1]. Conventional energy sources are also known to be one of the major causes of environmental pollution and global warming by emitting a wide variety of greenhouse gases (GHGs) [2]. GHG emissions can also pose a major risk to public health as well as the perceived quality of life [3]. Global efforts in promoting sustainability by The World Commission on Environmental and Development report in 1987 have led to an increased awareness of the adverse effects of using fossil fuels and the benefits of sustainability [4]. That increase in awareness has led to an increase in sustainability and sustainable-development-related research in a variety of fields.

Effective sustainable development depends on the ability to select appropriate sustainable development projects to ensure that the desired results are achieved. The viability of different project proposals, as well as limited resources available, must be considered carefully based on established criteria [5]. The selection process also includes considering many different criteria of the different project alternatives in an effort to determine the best possible project that can meet the desired goals. By ranking these key sustainable project characteristics or criteria, it helps project managers and decision makers focus on more important areas when evaluating the different project alternatives in addition to resource allocation.

The project selection process considers several different project factors or criteria as well as project goals and objectives [6]. This process usually takes place in a highly uncertain and complex environment. These uncertainties may be the result of unquantifiable measures or subjective judgments of experts about the relative importance of the different criteria used in the decision-making process [7]. The analytical hierarchy process (AHP) is one of the most commonly used techniques for project selection and assigns weights to different project factors used in the selection process. However, despite a recognition of the presence of uncertainty and ambiguity, AHP does not count for the ambiguity and uncertainty associated with project selection in an effective way [8]. To solve this problem, a combination of fuzzy numbers and AHP, known as the fuzzy analytic hierarchy process (FAHP), is used to account for the uncertainty and ambiguity in expert judgments [9].

The use of FAHP in sustainable project selection has mostly focused on evaluating different sustainable technology alternatives, with an emphasis on the technical aspects of these technologies, not necessarily the projects as a whole. This research improves the selection process of sustainable projects by developing a selection tool that considers the often-neglected criteria in the FAHP literature of novelty, uncertainty, team skill and experience, and technology information transfer, as they are described by Alyamani et al. [10], in addition to project cost. Accordingly, fuzzy AHP is used in this selection tool to rank these five selection criteria based on importance in the context of sustainable projects using input data from sustainable project experts. This tool will help project managers and decision makers focus on the selection criteria with higher weights of importance when evaluating different sustainable project alternatives. In addition, given that the selection criteria chosen for this research are not limited to the evaluation of a specific type of sustainable projects or a specific location, they can be used to evaluate different types of sustainable projects in different environments and locations.

This research is organized into five sections as follows: After the introduction section, Section 2 provides a literature review of relevant literature as well as major gaps found. Section 3 includes an explanation of the FAHP methodology and how it is implemented in this research to generate the results. Section 4 includes a discussion of the ranking results obtained from implementing the FAHP methodology and their relation to some of the existing literature. The final section (Section 5) of this research presents the conclusion, limitations, and future work.

2. Literature Review

Fuzzy AHP has been used in the literature by researchers in many different fields including project selection by assigning weights to selected project characteristics or criteria based on importance [11]. Bilgen and Şen [12] used a fuzzy AHP to develop a selection tool for six sigma projects. Their selection tool used resources, benefits, and effects as the major characteristics for their FAHP project selection tool. Enea and Piazza [6] used fuzzy AHP to develop a project selection tool based on the following characteristics: risk, cost, impact, and duration. Nguyen and Tran [13] studied the use of fuzzy AHP in construction projects for site selection, contractor selection, construction methods, risk assessment, and other areas related to construction projects. Other examples exist in the literature utilizing the fuzzy AHP methodology in project selection [14–16].

Fuzzy AHP has been used as part of sustainability and sustainable development research in recent years [11] across a broad spectrum of examples. Sabaghi et al. [17] used fuzzy AHP to evaluate product and process sustainability. FAHP was used in their research to assign weights to determine the importance of different economic, social, and environmental indicators in product development. Lespier et al. [7] used fuzzy AHP to quantify and rank key environmental impact criteria in maritime transportation systems (MTS) in an effort to help decision makers improve environmental sustainability in Maritime shipping. Ligus [8] utilized FAHP to evaluate sustainability in the development of different energy technologies based on determined economic, social, and environmental criteria. Li et al. [9] developed a fuzzy AHP based tool to evaluate the carbon performance of public projects by ranking different carbon emission criteria related to the design, construction, and operation phases of these projects. Other examples of using FAHP to rank the different economic, social, and environmental impacts of sustainable technologies also exist [18,19]. Malik et al. [20] provide a ranking for the following five sustainable project characteristics: technology, economic impact, environmental impact, planning time, and policy to aid in the selection between alternative sustainable projects in Oman. However, since the standard AHP methodology was used to rank these characteristics, the uncertainty in experts' subjective judgments was not considered.

Although previous research demonstrates the use of FAHP to evaluate sustainability and sustainable project development, the focus has mainly been on the selection between different sustainable technology alternatives not necessarily the projects as a whole with an emphasis on the technical aspects of these technologies such as technology efficiency, reliability, scalability, and many other technical aspects in addition to the economic, social, and environmental impacts of these technologies [11]. Even though these technical factors and the impacts of these technologies are important to consider when selecting from different sustainable project alternatives, it is also important to consider the characteristics of these projects as a whole in the selection process not just the sustainable technologies used and their impact. More specifically, there seems to be little research in the FAHP literature that combines project cost and the more neglected, but crucial, project selection criteria of novelty, uncertainty, skill and experience, and technology information transfer and ranking them based on importance in the context of sustainable projects. These criteria can be used to evaluate sustainable projects as a whole regardless of the type of sustainable technology used and location of these projects.

Research Question

This research aims to fill the gap in the literature discussed above and answer the following research question specifically:

- Among the five chosen sustainable project selection criteria in this research, which one of them is the most important to consider when selecting between different sustainable project alternatives?

Given that novelty, uncertainty, team skill and experience, technology information transfer, and project cost are considered universal key criteria used to evaluate sustainable projects [10,21], the results from this research will provide project managers and decision makers presented with multiple sustainable project alternatives with a globally applicable selection tool capable of identifying the most important selection criteria when presented with multiple sustainable project alternatives.

3. Methodology

Project selection is an increasingly complicated process. This is due to the many interrelated variables that are used to evaluate these projects. Each of these variables has potential consequences to the project that must be determined to ensure the success of the project. In addition, the uncertainties surrounding both measuring these variables and determining their consequences on the project can be significant. These uncertainties sometimes stem from information that is difficult to quantify, or from subjective opinions of decision makers [7]. Such uncertainties make the project selection process highly subjective and at risk of inaccurate information and judgments. This results in a lack of consensus on the relative importance of the different criteria used to evaluate projects in the selection process [6].

3.1. *Fuzzy AHP and Fuzzy Logic*

Multi-criteria decision-making (MCDM) techniques are extremely beneficial for project selection problems when considering different selection criteria. These techniques use mathematical models and simulations to aid in the project selection process. AHP, introduced by Saaty [22], is one of the most common and established MCDM techniques in project selection [15]. However, for these techniques to yield meaningful results, they need crisp and specific input data, which are usually difficult to obtain in project selection situations due to the subjective and uncertain nature of experts' judgments. Fuzzy AHP was developed to handle such uncertain and subjective input data more effectively than conventional MCDM techniques [7]. Fuzzy AHP applies the fuzzy set theory to allow researchers and decision makers to convert uncertain and vague linguistic input information from experts, such as the phrase "A lot more important", for example, to specific decisions intervals that are a lot more convenient to deal with by decision makers [15,23]. As project selection becomes increasingly global, this is a critical dimension to evaluate effectively.

The concept of fuzzy numbers used in the FAHP represents a range of possible values for a specific variable or rating. This means that a single ambiguous linguistic rating will be translated into a fuzzy number consisting of a range of numbers [24]. In fuzzy theory, it is more convenient to use triangular fuzzy numbers (TFNs) because of their computational simplicity and usefulness in representing information in a fuzzy environment [25]. TFNs are represented as three numbers (l, m, u) where the variables l, m, and u indicate the lowest possible value, the modal or most likely value, and the upper or highest possible value, respectively [7]. The mathematical representation of a fuzzy number A with a membership function $\mu_A(x)$ is depicted in Equation (1), as shown in Shukla et al. [24] and Hsieh et al. [26].

$$\mu_A(x) = \begin{cases} 0 & x < l; \\ \frac{x-l}{m-l} & l \leq x \leq m; \\ \frac{u-x}{u-m} & m \leq x \leq u; \\ 0 & x > u. \end{cases} \quad (1)$$

The geometric representation of the fuzzy number A from Equation (1) is shown in Figure 1, adapted from Lespier et al. [7] and Sun [27].

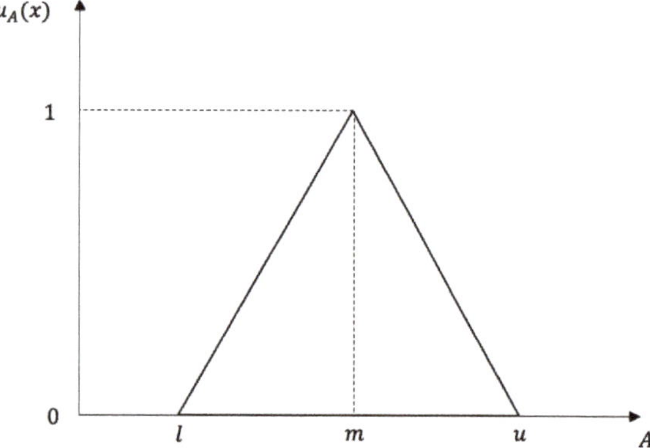

Figure 1. A triangular fuzzy number (TFN), A [7,27].

3.2. FAHP Selection Criteria

Alyamani and Long [21] and Alyamani et al. [10] identified four common key project characteristics that are used to evaluate sustainable projects in different institutional environments. This research extends their work by utilizing the characteristics they identified in addition to project cost as a fifth characteristic. The five characteristics are then used as selection criteria in evaluating multiple sustainable project alternatives. Using these characteristics as selection criteria develops a selection tool that can be used to evaluate projects in different environments regardless of location. Consequently, this research aims to rank novelty, uncertainty, skill and experience, technology information transfer, and project cost from the context of sustainability as part of project selection in different environments and locations.

Novelty describes the degree to which a project differs from what is considered standard and established in terms of sustainable practices, processes, and technologies. In other words, this refers to the originality of the project and the maturity of the selected sustainable practices and technologies [28]. Undertaking a novel project that is utilizing completely new sustainable technologies or practices presents its own set of challenges and requires a certain level of resources and capabilities to ensure the successful implementation of such projects as opposed to more mature sustainable projects using standard and established sustainable practices and technologies [10,29].

Project uncertainty is generally defined in the literature as negative events for which both the consequence and probability of occurrence is unknown [30,31]. Different projects have different levels and sources of uncertainty [10]. In any case, however, these different sources of uncertainty, whether it be technological, financial, environmental, political, or any other source, should be outlined and addressed with appropriate mitigation plans to reduce their potential impact on the project should they occur.

The skill and experience criterion describes the level of skill and experience a project team is required to possess to be able to complete the project tasks effectively and efficiently, thus ensuring the successful completion of the project [10]. This criterion essentially addresses matching workforce capabilities with the project requirements [32]. Some sustainable projects require a highly skilled and experienced project team to be able to successfully complete the project, while other sustainable projects require relatively lower levels of skill and experience. The availability of the required workforce capabilities within the location of the evaluated project alternatives is an important component of this criterion. Project tasks can range from being trivial and standard all the way to complex and unusual. Consequently, choosing a project team with the appropriate know-how and sufficient level of experience to undertake these tasks and implement the chosen sustainable technology or practice is crucial in achieving project success and ensuring that project goals are met.

Technology information transfer, originally presented by Stock and Tatikonda [32], describes the amount of sustainable technology information being exchanged between the supplier of the sustainable technology and the project team implementing that technology. In other words, it describes the amount of interaction required between a supplier of a technology and the recipient of that technology to ensure the successful integration and implementation of said technology in the project. Selecting the appropriate technology and making sure it is correctly implemented in the project is one of the major steps towards achieving project goals. The level of information sharing between the two parties can vary significantly from project to project depending on the type of technology implemented. Stock and Tatikonda [32] explain that the level of information sharing between the supplier of the technology and the project team can range from a simple "arms-length" purchase requiring trivial information sharing, all the way to a "co-development" type of technology information sharing where both the supplier of the technology and the project team work closely together on the details of the design and specifications to ensure successful integration of the technology in the project [10].

Project cost essentially describes the total cost of the project including the initial investment cost and subsequent annual project costs. This criterion was added because it is considered one of the major driving factors in sustainable development and sustainable project selection [11]. One of the major challenges facing sustainable energy projects is competing with conventional energy sources in financial cost. However, the reduction in sustainable development costs in recent years in addition to the consideration of the indirect costs associated with conventional energy sources has somewhat balanced the scales between sustainable and conventional energy sources from the economic perspective [20]. Nonetheless, the costs associated with sustainable energy development in the international stage remain one of the major driving forces in sustainable energy project development.

A summary of the criteria explained above and their notations as used in this research are presented in Table 1.

Table 1. Key sustainable project selection criteria used in fuzzy analytic hierarchy process (FAHP).

Notation	Project Selection Criteria
C1	Project Cost
C2	Novelty
C3	Uncertainty
C4	Skill and Experience
C5	Technology Information Transfer

Based on these criteria, a typical hierarchy model of the sustainable project selection process is created, as shown in Figure 2, which consists of three levels: the goal of evaluating sustainable project alternatives, the criteria used to evaluate these alternatives as presented in Table 1, and the sustainable project alternatives to be evaluated using these criteria. As such, the prioritization of weights for the presented criteria using fuzzy analytic hierarchy process (FAHP) will aid in the selection process when presented with different sustainable project alternatives.

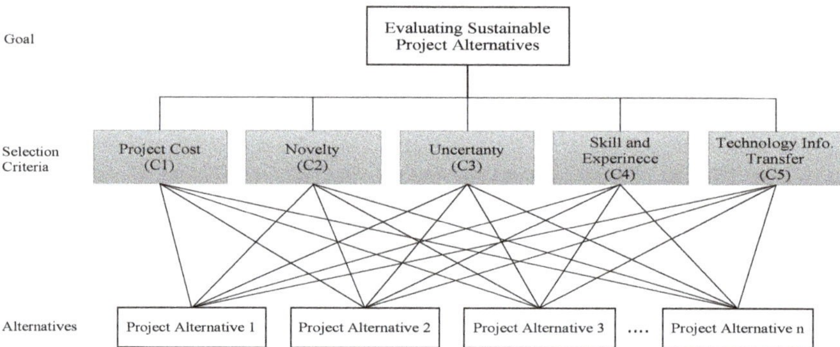

Figure 2. The hierarchy model for sustainable project selection.

3.3. The Application of FAHP for Weight Calculation

After defining the five sustainable project criteria, as shown in the previous subsection, the first step in determining the priority weights of these criteria is collecting the opinions of experts in sustainability and sustainable development regarding the relative importance of these criteria in sustainable project selection. In this research, a number of literature publications related to sustainable project selection and sustainable development as well as some prominent project management literature covering the chosen criteria were selected and evaluated, as part of the literature review for this research, to serve as the voice of experts in determining preferences among the five different criteria shown in Table 1. These studies were closely reviewed in an effort to determine the relative importance of these criteria and preference patterns, as presented by the authors of these publications. The list of the chosen literature publications is shown in Table 2.

Table 2. Selected expert literature used for the evaluation of criteria.

Expert	Source(s)
E1	Malik et al. [20]
E2	Alyamani et al. [10]
E3	Sabaghi et al. [17]
E4	Shenhar and Dvir [29], Stock and Tatikonda [32]
E5	Chen et al. [33]
E6	Wang et al. [28]
E7	Işik and Aladağ [34]
E8	Hatefi and Tamošaitienė [16]
E9	Luthra et al. [35]
E10	Solangi et al. [36]

The second step in determining the priority weights of the five sustainable project criteria is utilizing the expert opinions from the literature in Table 2 based on the linguistic variables and triangular fuzzy numbers (TFNs), shown in Table 3, as presented by Ballı and Korukoğlu [25]. In this step, expert opinions are gathered from the literature and translated into the linguistic variables. After creating the pairwise comparison matrix representing the opinions of each of the ten experts shown in Table 1 using the linguistic variables, these ten matrices are then combined to form the combined pairwise comparison matrix shown in Table 4.

Table 3. Linguistic variables and triangular fuzzy number scale.

Linguistic Variable	Triangular Fuzzy Numbers (TFN)	Reciprocal TFNs
Equally Important (E)	(1, 1, 1)	(1, 1, 1)
Weakly Important (W)	(1, 3, 5)	(1/5, 1/3, 1)
Fairly Important (F)	(3, 5, 7)	(1/7, 1/5, 1/3)
Strongly Important (S)	(5, 7. 9)	(1/9, 1/7, 1/5)
Absolutely Important (A)	(7, 9, 11)	(1/11, 1/9, 1/7)

Source: adapted from Ballı and Korukoğlu [25].

Table 4. Pairwise comparison matrix using linguistic variables.

Criteria	Expert	C1	C2	C3	C4	C5
C1	**E1**	E	F	S	A	A
	E2	E	S^{-1}	S^{-1}	F	S
	E3	E	S	F	F	A
	E4	E	S^{-1}	S^{-1}	F	W
	E5	E	S	F	F	A
	E6	E	A	F	A	F
	E7	E	F	A	F	S
	E8	E	F	W^{-1}	S	W^{-1}
	E9	E	F	A	S	F
	E10	E	W	F	S	A

Table 4. *Cont.*

Criteria	Expert	C1	C2	C3	C4	C5
C2	**E1**	F^{-1}	E	E	S	A
	E2	S	E	W	S	A
	E3	S^{-1}	E	F^{-1}	F^{-1}	W
	E4	S	E	W	A	S
	E5	S^{-1}	E	F^{-1}	S^{-1}	W
	E6	A^{-1}	E	S^{-1}	W	S^{-1}
	E7	F^{-1}	E	S	E	F
	E8	F^{-1}	E	S^{-1}	F	S^{-1}
	E9	F^{-1}	E	S	F	W^{-1}
	E10	W^{-1}	E	F	S	S
C3	**E1**	S^{-1}	E^{-1}	E	F	S
	E2	S	W^{-1}	E	S	S
	E3	F^{-1}	F	E	W	S
	E4	S	W^{-1}	E	A	S
	E5	F^{-1}	F	E	F^{-1}	F
	E6	F^{-1}	S	E	S	W^{-1}
	E7	A^{-1}	S^{-1}	E	S^{-1}	F^{-1}
	E8	W	S	E	A	E
	E9	A^{-1}	S^{-1}	E	F^{-1}	S^{-1}
	E10	F^{-1}	F^{-1}	E	F	S
C4	**E1**	A^{-1}	S^{-1}	F^{-1}	E	W
	E2	F^{-1}	S^{-1}	S^{-1}	E	W
	E3	F^{-1}	F	W^{-1}	E	S
	E4	F^{-1}	A^{-1}	A^{-1}	E	F^{-1}
	E5	F^{-1}	S	F	E	S
	E6	A^{-1}	W^{-1}	S^{-1}	E	S^{-1}
	E7	F^{-1}	E^{-1}	S	E	F
	E8	S^{-1}	F^{-1}	A^{-1}	E	A^{-1}
	E9	S^{-1}	F^{-1}	F	E	F^{-1}
	E10	S^{-1}	S^{-1}	F^{-1}	E	W
C5	**E1**	A^{-1}	A^{-1}	S^{-1}	W^{-1}	E
	E2	S^{-1}	A^{-1}	S^{-1}	W^{-1}	E
	E3	A^{-1}	W^{-1}	S^{-1}	S^{-1}	E
	E4	W^{-1}	S^{-1}	S^{-1}	F	E
	E5	A^{-1}	W^{-1}	F^{-1}	S^{-1}	E
	E6	F^{-1}	S	W	S	E
	E7	S^{-1}	F^{-1}	F	F^{-1}	E
	E8	W	S	E^{-1}	A	E
	E9	F^{-1}	W	S	F	E
	E10	A^{-1}	S^{-1}	S^{-1}	W^{-1}	E

These linguistic variables in the combined matrix are then further translated into the corresponding triangular fuzzy numbers (TFNs) and reciprocal TFNs based on the scale shown in Table 3, resulting in the combined TFN pairwise comparison matrix, shown in Table 5.

Table 5. Pairwise comparison matrix using TFNs.

Criteria	Expert	C1	C2	C3	C4	C5
C1	E1	(1, 1, 1)	(3, 5, 7)	(5, 7. 9)	(7, 9, 11)	(7, 9, 11)
	E2	(1, 1, 1)	(1/9, 1/7, 1/5)	(1/11, 1/9, 1/7)	(3, 5, 7)	(5, 7. 9)
	E3	(1, 1, 1)	(5, 7. 9)	(3, 5, 7)	(3, 5, 7)	(7, 9, 11)
	E4	(1, 1, 1)	(1/9, 1/7, 1/5)	(1/9, 1/7, 1/5)	(3, 5, 7)	(1, 3, 5)
	E5	(1, 1, 1)	(5, 7. 9)	(3, 5, 7)	(3, 5, 7)	(7, 9, 11)
	E6	(1, 1, 1)	(7, 9, 11)	(3, 5, 7)	(7, 9, 11)	(3, 5, 7)
	E7	(1, 1, 1)	(3, 5, 7)	(7, 9, 11)	(3, 5, 7)	(5, 7. 9)
	E8	(1, 1, 1)	(3, 5, 7)	(1/5, 1/3, 1)	(5, 7. 9)	(1/5, 1/3, 1)
	E9	(1, 1, 1)	(3, 5, 7)	(7, 9, 11)	(5, 7. 9)	(3, 5, 7)
	E10	(1, 1, 1)	(1, 3, 5)	(3, 5, 7)	(5, 7. 9)	(7, 9, 11)
C2	E1	(1/7, 1/5, 1/3)	(1, 1, 1)	(1, 1, 1)	(5, 7. 9)	(7, 9, 11)
	E2	(5, 7. 9)	(1, 1, 1)	(1, 3, 5)	(5, 7. 9)	(7, 9, 11)
	E3	(1/9, 1/7, 1/5)	(1, 1, 1)	(1/7, 1/5, 1/3)	(1/7, 1/5, 1/3)	(1, 3, 5)
	E4	(5, 7. 9)	(1, 1, 1)	(1, 3, 5)	(7, 9, 11)	(5, 7. 9)
	E5	(1/9, 1/7, 1/5)	(1, 1, 1)	(1/7, 1/5, 1/3)	(1/9, 1/7, 1/5)	(1, 3, 5)
	E6	(1/11, 1/9, 1/7)	(1, 1, 1)	(1/9, 1/7, 1/5)	(1, 3, 5)	(1/9, 1/7, 1/5)
	E7	(1/7, 1/5, 1/3)	(1, 1, 1)	(5, 7. 9)	(1, 1, 1)	(3, 5, 7)
	E8	(1/7, 1/5, 1/3)	(1, 1, 1)	(1/9, 1/7, 1/5)	(3, 5, 7)	(1/9, 1/7, 1/5)
	E9	(1/7, 1/5, 1/3)	(1, 1, 1)	(5, 7. 9)	(3, 5, 7)	(1/5, 1/3, 1)
	E10	(1/5, 1/3, 1)	(1, 1, 1)	(3, 5, 7)	(5, 7. 9)	(5, 7. 9)
C3	E1	(1/9, 1/7, 1/5)	(1, 1, 1)	(1, 1, 1)	(3, 5, 7)	(5, 7. 9)
	E2	(5, 7. 9)	(1/5, 1/3, 1)	(1, 1, 1)	(5, 7. 9)	(5, 7. 9)
	E3	(1/7, 1/5, 1/3)	(3, 5, 7)	(1, 1, 1)	(1, 3, 5)	(5, 7. 9)
	E4	(5, 7. 9)	(1/5, 1/3, 1)	(1, 1, 1)	(7, 9, 11)	(5, 7. 9)
	E5	(1/7, 1/5, 1/3)	(3, 5, 7)	(1, 1, 1)	(1/7, 1/5, 1/3)	(3, 5, 7)
	E6	(1/7, 1/5, 1/3)	(5, 7. 9)	(1, 1, 1)	(5, 7. 9)	(1/5, 1/3, 1)
	E7	(1/11, 1/9, 1/7)	(1/9, 1/7, 1/5)	(1, 1, 1)	(1/9, 1/7, 1/5)	(1/7, 1/5, 1/3)
	E8	(1, 3, 5)	(5, 7. 9)	(1, 1, 1)	(7, 9, 11)	(1, 1, 1)
	E9	(1/11, 1/9, 1/7)	(1/9, 1/7, 1/5)	(1, 1, 1)	(1/7, 1/5, 1/3)	(1/9, 1/7, 1/5)
	E10	(1/7, 1/5, 1/3)	(1/7, 1/5, 1/3)	(1, 1, 1)	(3, 5, 7)	(5, 7. 9)
C4	E1	(1/11, 1/9, 1/7)	(1/9, 1/7, 1/5)	(1/7, 1/5, 1/3)	(1, 1, 1)	(1, 3, 5)
	E2	(1/7, 1/5, 1/3)	(1/9, 1/7, 1/5)	(1/9, 1/7, 1/5)	(1, 1, 1)	(1, 3, 5)
	E3	(1/7, 1/5, 1/3)	(3, 5, 7)	(1/5, 1/3, 1)	(1, 1, 1)	(5, 7. 9)
	E4	(1/7, 1/5, 1/3)	(1/11, 1/9, 1/7)	(1/11, 1/9, 1/7)	(1, 1, 1)	(1/7, 1/5, 1/3)
	E5	(1/7, 1/5, 1/3)	(5, 7. 9)	(3, 5, 7)	(1, 1, 1)	(5, 7. 9)
	E6	(1/11, 1/9, 1/7)	(1/5, 1/3, 1)	(1/9, 1/7, 1/5)	(1, 1, 1)	(1/9, 1/7, 1/5)
	E7	(1/7, 1/5, 1/3)	(1, 1, 1)	(5, 7. 9)	(1, 1, 1)	(3, 5, 7)

Table 5. *Cont.*

Criteria	Expert	C1	C2	C3	C4	C5
	E8	(1/9, 1/7, 1/5)	(1/7, 1/5, 1/3)	(1/11, 1/9, 1/7)	(1, 1, 1)	(1/11, 1/9, 1/7)
	E9	(1/9, 1/7, 1/5)	(1/7, 1/5, 1/3)	(3, 5, 7)	(1, 1, 1)	(1/7, 1/5, 1/3)
	E10	(1/9, 1/7, 1/5)	(1/9, 1/7, 1/5)	(1/7, 1/5, 1/3)	(1, 1, 1)	(1, 3, 5)
C5	E1	(1/11, 1/9, 1/7)	(1/11, 1/9, 1/7)	(1/9, 1/7, 1/5)	(1/5, 1/3, 1)	(1, 1, 1)
	E2	(1/9, 1/7, 1/5)	(1/11, 1/9, 1/7)	(1/9, 1/7, 1/5)	(1/5, 1/3, 1)	(1, 1, 1)
	E3	(1/11, 1/9, 1/7)	(1/5, 1/3, 1)	(1/9, 1/7, 1/5)	(1/9, 1/7, 1/5)	(1, 1, 1)
	E4	(1/5, 1/3, 1)	(1/9, 1/7, 1/5)	(1/9, 1/7, 1/5)	(3, 5, 7)	(1, 1, 1)
	E5	(1/11, 1/9, 1/7)	(1/5, 1/3, 1)	(1/7, 1/5, 1/3)	(1/9, 1/7, 1/5)	(1, 1, 1)
	E6	(1/7, 1/5, 1/3)	(5, 7. 9)	(1, 3, 5)	(5, 7. 9)	(1, 1, 1)
	E7	(1/9, 1/7, 1/5)	(1/7, 1/5, 1/3)	(3, 5, 7)	(1/7, 1/5, 1/3)	(1, 1, 1)
	E8	(1, 3, 5)	(5, 7. 9)	(1, 1, 1)	(7, 9, 11)	(1, 1, 1)
	E9	(1/7, 1/5, 1/3)	(1, 3, 5)	(5, 7. 9)	(3, 5, 7)	(1, 1, 1)
	E10	(1/11, 1/9, 1/7)	(1/9, 1/7, 1/5)	(1/9, 1/7, 1/5)	(1/5, 1/3, 1)	(1, 1, 1)

Once the TFN pairwise comparison matrix is created, as shown above, it can be used to calculate the weight of importance for the five criteria. This calculation is performed in three main steps. The first step is to combine the fuzzy pairwise comparison from all ten experts for each of the five criteria. This can be done by calculating the geometric mean of the experts' opinions. To calculate the fuzzy geometric mean, the geometric mean method introduced by Buckley [37] is used leading to the fuzzy geometric mean pairwise comparison matrix shown in Table 6.

Table 6. Fuzzy geometric mean pairwise comparison matrix.

Criteria	C1	C2	C3	C4	C5
C1	(1, 1, 1)	(1.676, 2.647, 3.657)	(1.446, 2.125, 3.071)	(4.143, 6.221, 8.262)	(3.187, 4.904, 7.020)
C2	(0.273, 0.378, 0.597)	(1, 1, 1)	(0.672, 1.061, 1.513)	(1.621, 2.410, 3.249)	(1.247, 2.034, 3.045)
C3	(0.315, 0.459, 0.678)	(0.661, 0.943, 1.487)	(1, 1, 1)	(1.380, 2.104, 2.970)	(1.404, 1.951, 2.780)
C4	(0.121, 0.161, 0.241)	(0.308, 0.415, 0.617)	(0.337, 0.475, 0.725)	(1, 1, 1)	(0.659, 1.154, 1.719)
C5	(0.142, 0.204, 0.314)	(0.328, 0.492, 0.802)	(0.360, 0.512, 0.712)	(0.582, 0.866, 1.517)	(1, 1, 1)

The second step in calculating the criteria weights of importance is determining the fuzzy relative importance weight or the fuzzy synthetic extent of each of the five criteria. To do that, the extent analysis method introduced by Chang [38] is applied in this research, as shown in Equations (2–5). Let $G = \{g_1, g_2, g_3, \ldots, g_n\}$ be a goal set. Each criterion is taken and the extent analysis for each goal g_i is performed, respectively [25,39]. Accordingly, the m extent value for each criterion is obtained as follows: $M_{g_i}^1, M_{g_i}^2, M_{g_i}^3, \ldots, M_{g_i}^m$, where g_i $(i = 1,2,3,\ldots,n)$ is the goal set and $M_{g_i}^j$ $(j = 1,2,3,\ldots, m)$ are all TFNs. The value of the fuzzy synthetic extent (S_i) with respect to the ith criterion is defined as shown in Equation (2).

$$S_i = \sum_{j=1}^{m} M_{g_i}^j \otimes \left[\sum_{i=1}^{n} \sum_{j=1}^{m} M_{g_i}^j \right]^{-1} \quad (2)$$

In order to calculate $\sum\limits_{j=1}^{m} M_{g_i}^{j}$, a fuzzy addition operation of the m extent is used for a certain matrix, as shown in Equation (3). This can be done following the addition of the fuzzy number process shown in Sun [27].

$$\sum_{j=1}^{m} M_{g_i}^{j} = \left(\sum_{j=1}^{m} l_j, \sum_{j=1}^{m} m_j, \sum_{j=1}^{m} u_j \right) \tag{3}$$

where the variables l, m, and u indicate the lowest possible value, the modal or most likely value, and the upper or highest possible value, respectively, as explained earlier in this research. The next logical operation is to calculate $\sum\limits_{i=1}^{n} \sum\limits_{j=1}^{m} M_{g_i}^{j}$ by performing another fuzzy addition operation of $M_{g_i}^{j}$ $(j = 1, 2, 3, \ldots, m)$, as shown in Equation (4).

$$\sum_{i=1}^{n} \sum_{j=1}^{m} M_{g_i}^{j} = \left(\sum_{i=1}^{n} l_i, \sum_{i=1}^{n} m_i, \sum_{i=1}^{n} u_i \right) \tag{4}$$

Finally, $\left[\sum\limits_{i=1}^{n} \sum\limits_{j=1}^{m} M_{g_i}^{j} \right]^{-1}$ is determined by calculating the inverse of the vector above as shown in Equation (5).

$$\left[\sum_{i=1}^{n} \sum_{j=1}^{m} M_{g_i}^{j} \right]^{-1} = \left(\frac{1}{\sum_{i=1}^{n} u_i}, \frac{1}{\sum_{i=1}^{n} m_i}, \frac{1}{\sum_{i=1}^{n} l_i} \right) \tag{5}$$

Equations (2)–(5) are now applied to the TFNs obtained in this research. To determine the fuzzy synthetic extent to the criteria chosen in this research, the $\sum_{j=1}^{m} M_{g_i}^{j}$ value is first calculated for each row of the matrix shown in Table 6. For example, for C1:

$$C1 = (1 + 1.676 + 1.446 + 4.143 + 3.187, 1 + 2.647 + 2.125 + 6.221 + 4.904, 1 + 3.657 + 3.071 + 8.262 + 7.020)$$

$$C1 = (11.452, 16.897, 23.010)$$

Accordingly, the $\sum_{i=1}^{n} \sum_{j=1}^{m} M_{g_i}^{j}$ value is calculated for each of the five criteria in Table 6 by applying Equation (4) as follows:

$$\begin{aligned} \textstyle\sum_{i=1}^{n} \sum_{j=1}^{m} M_{g_i}^{j} &= (11.452,\ 16.897,\ 23.010) \oplus (4.813,\ 6.883,\ 9.404) \oplus (4.760,\ 6.457,\ 8.915) \\ &\oplus (2.425,\ 3.205,\ 4.302) \oplus (2.412,\ 3.074,\ 4.345) \\ &= (25.862, 36.516, 49.976) \end{aligned}$$

Based on that, the reciprocal value $\left[\sum_{i=1}^{n} \sum_{j=1}^{m} M_{g_i}^{j} \right]^{-1}$ is calculated by applying Equation (5) as follows:

$$\left[\sum_{i=1}^{n} \sum_{j=1}^{m} M_{g_i}^{j} \right]^{-1} = \left(\frac{1}{49.976}, \frac{1}{36.516}, \frac{1}{25.862} \right) = (0.020,\ 0.027,\ 0.039)$$

Finally, the value of the fuzzy synthetic extent (S_i) with respect to the ith criterion is calculated for each criterion, as shown in Equation (2). For example, the value of the fuzzy synthetic extent for the first criterion S_1 is calculated as follows:

$$S_1 = (11.452,\ 16.897,\ 23.010) \otimes (0.020,\ 0.027,\ 0.039) = (0.229,\ 0.436,\ 0.893)$$

The fuzzy synthetic extent or the fuzzy relative importance weights resulting from applying the same process to the remaining criteria is presented in Table 7.

Table 7. Fuzzy synthetic extent of sustainable project selection criteria.

Criteria	S_i Low	S_i Med	S_i Upper
C1	0.229	0.463	0.893
C2	0.096	0.188	0.364
C3	0.095	0.177	0.345
C4	0.049	0.088	0.166
C5	0.048	0.084	0.168

The third and final step in calculating the criteria weights of importance is the defuzzification of the fuzzy criteria weights shown in Table 7. To defuzzify these weights, the defuzzification method shown in Equation (6), as presented in Sun [27] and Lespier et al. [7], is used to obtain the best non-fuzzy priority (BNP) or crisp weights of the criteria.

$$BNP_{S_i} = \frac{\left[\left(u_{s_i} - l_{s_i}\right) + \left(m_{s_i} - l_{s_i}\right)\right]}{3} + l_{s_i} \qquad where\ i = 1, 2, \ldots, 5 \tag{6}$$

As an example, applying Equation (6) to calculate the BNP for criterion 1 is done as follows:

$$BNP_{S_1} = \frac{[(0.893 - 0.229) + (0.463 - 0.229)]}{3} + 0.229 = 0.528$$

Accordingly, the crisp weights for the remaining criteria are calculated. Using these BNP values, the criteria can be ranked based on importance, where the criterion with the highest BNP is set as the most important, while the criterion with the lowest BNP is set as the least important, as shown in Table 8.

Table 8. Best non-fuzzy priority (BNP) or crisp criteria weights.

Criteria	BNP	Rank
C1—Project Cost	0.528	1
C2—Novelty	0.216	2
C3—Uncertainty	0.206	3
C4—Skill and Experience	0.101	4
C5—Technology Info. Transfer	0.100	5

4. Discussion of Results

Sustainable project selection is an important step in successful sustainable development. Selecting the appropriate sustainable project is a major step in ensuring the success of the project and, thus, achieving the desired sustainability and project goals. The sustainable project selection process depends on a wide variety of criteria. One of the major challenges facing decision makers in sustainable project selection is the strong dependence on the subjective judgments of experts in prioritizing the project selection criteria, as well as the uncertainties associated with these subjective judgments. To help overcome these challenges, a fuzzy multi-criteria decision-making methodology has been implemented in this research. FAHP has been used in this research to rank five key sustainable project selection criteria shown in Table 1 by calculating the relative weight of importance for each of these selection criteria.

The results show that the most important criterion to consider in sustainable project selection is project cost (C1) with an importance weight (BNP) of 0.528. This mainly includes different sources of cost for the project such as the project's initial investment cost, maintenance cost, labor cost, operating costs, and any other cost associated with the project over its life cycle that can differ from one location or country to the other [28]. This result has been mostly consistent with what has been shown in the literature when considering the economic aspect of sustainable projects. As mentioned earlier in this research, project cost has been one of the major factors influencing sustainable development in the international stage due to concerns that renewable and sustainable energy projects cannot compete economically with conventional energy projects [20]. The different sources of project cost including the investment cost, operating and maintenance costs, and labor costs are also considered as variables in the measurement of project efficiency that can be used to evaluate sustainable projects, as shown by Švajlenka and Kozlovská [40].

The second and third most important criteria to consider in sustainable project selection in this research are novelty (C2) and uncertainty (C3) with BNPs of 0.216 and 0.206, respectively. Both of these criteria are also considered one of the most important in sustainable project selection. As mentioned earlier in this research, novelty mainly focuses on the originality and maturity of the sustainable technologies and practices used in these projects. It is also an indicator of how widespread a sustainable technology or practice is in the location or country these projects exist in and the improvement potential of these technologies and practices [28]. The novelty of the sustainable technologies and practices used in projects can also potentially help accelerate the opportunities for sustainability adoption in communities [33]. Uncertainty can include different sub criteria that can be on both a local or international scale such as financial uncertainty, technological uncertainty, environmental uncertainty, and political uncertainty each with a different impact on sustainable projects. Since most of the sustainable project selection literature focus on the technical aspect of sustainable technologies, there has been an emphasis on the technical uncertainties associated with these technologies. Nonetheless, other international or local sources of uncertainty are also important and should also be considered just as crucial in sustainable project selection, since they can potentially hinder the use of sustainable technologies and practices in a given location [35].

The two least important criteria out of the five considered in this research based on the selected experts' opinions are skill and experience (C4) and technology information transfer (C5) with BNPs of 0.101 and 0.100, respectively. These results show that both criteria have a relatively similar level of importance with skill and experience being just slightly more important than technology information transfer. However, these results cannot be interpreted as implying that these two criteria are not important and should not be considered in the selection of sustainable projects. They simply mean that the selected experts prioritize the other three criteria over skill and experience and technology information transfer when selecting between different sustainable project alternatives.

As explained earlier in this research, skill and experience refers to having the appropriate know-how to successfully undertake a selected sustainable project. Kahraman et al. [41], Amer and Daim [42], and Solangi et al. [36] all argue that having the appropriate human resources with the required skills and experience to build, operate, and maintain the sustainable project in the location or country in which these projects exist is a crucial factor to consider when selecting between different sustainable project alternatives to ensure the success of the project. Technology information transfer refers to the level of technology information sharing or communication between a supplier of a technology and the project team implementing that technology. The unavailability of the adequate technological information in a specific location or country as well as inadequate information sharing and communication may be considered as one of the greatest barriers to successful sustainable technology implementation and, ultimately, sustainable project success [35]. This information can include sustainable technology specifications, design, materials used, or any other technology information that is crucial to successful project implementation and, thus, achieving the overall goals of the project. For example, Švajlenka et al. [43] emphasized the importance of considering such information as environmental parameters in improving the decision-making process when evaluating the different project alternatives to examine whether or not these projects would meet the overall sustainable goals.

The selection criteria chosen for this research are not limited to the evaluation of a specific type of sustainable projects or a specific location. Instead, these criteria are applicable to evaluate different types of sustainable projects in different environments and geographical locations [10]. Moreover, one of the major benefits of using FAHP to rank these criteria based on a number of diverse sources of expert opinions is that it is designed to minimize any uncertainty or biases that are associated with the subjective judgments of these experts when performing the pairwise comparison [44,45]. Accordingly, the results presented in this research reflect the consensus among these diverse expert sources regarding the relative importance of the selection criteria regardless of any subjective judgment or biases.

5. Conclusions

This research implements the fuzzy analytic hierarchy process (FAHP) methodology as a multi-criteria decision-making (MCDM) approach to develop a sustainable project selection tool that quantifies and ranks five key sustainable project criteria based on importance. This selection tool can be applied by any project manager or decision maker when evaluating different sustainable project alternatives for selection regardless of the type, environment, and location of these projects. The criteria chosen in this research are novelty, uncertainty, team skill and experience, technology information transfer, and project cost. Prioritizing these criteria based on relative importance helps project managers and decision makers identify more important project elements that require additional attention, better allocate resources, as well as improve the selection process when evaluating different sustainable project alternatives. This research utilizes the existing literature examined as part of the literature review process to represent the voice of experts on the relative importance of the selected criteria.

The results from the FAHP methodology in this research answers the research question introduced earlier by showing that project cost is the most important criterion to consider when evaluating different sustainable project alternatives with a best non-fuzzy priority (BNP) of 0.528. This indicates that sustainable development is still significantly driven by economic factors specific to location. The second and third most important criteria to consider in sustainable project selection based on the FAHP results are novelty and uncertainty with BNPs of 0.216 and 0.206, respectively. This indicates that the originality and maturity of the sustainable technologies and practices used in these projects, as well as the different sources of uncertainty surrounding such projects, are also strong driving factors in sustainable project selection. Finally, the FAHP results show that the two least important criteria out of the five considered in this research are skill and experience and technology information transfer with BNPs of 0.101 and 0.100, respectively. This represents possible good news for developing economies that should be considered as part of future research.

The limitations associated with this research include the small sample size of literature considered to act as the voice of experts in the pairwise comparison of the chosen criteria. A larger sample size in the future could yield more accurate results regarding the relative importance of the selected criteria. It is also important to note that these results are limited to the knowledge and experiences of the chosen experts. Another potential limitation of this research is the use of literature to act as the voice of experts. This could add another layer of uncertainty and subjective judgment that stems from the interpretations and opinions of the researchers utilizing the literature, which is not accounted for by the FAHP. Future research should focus on gathering input data from sustainable project researchers and practitioners in an effort to gather direct input and, thus, eliminating any need for interpretation by the researchers.

Author Contributions: Conceptualization, R.A. and S.L.; data curation, R.A.; formal analysis, R.A.; investigation, R.A.; methodology, R.A.; project administration, S.L.; resources, R.A.; software, R.A.; supervision, S.L.; validation, R.A.; visualization, R.A.; writing—original draft, R.A.; writing—review and editing, S.L. All authors have read and agreed to the published version of the manuscript.

Funding: This research is partially funded by the Saudi Arabian Cultural Mission (SACM) and the Engineering Management & Systems Engineering Department at Missouri University of Science & Technology. The authors, therefore, gratefully acknowledge SACM and the Engineering Management & Systems Engineering Department at Missouri S&T for their financial support.

Conflicts of Interest: The authors have declared no conflict of interest.

References

1. Qin, R.; Grasman, S.E.; Long, S.; Lin, Y.; Thomas, M. A framework of cost-effectiveness analysis for alternative energy strategies. *Eng. Manag. J.* **2012**, *24*, 18–35. [CrossRef]
2. Almasoud, A.; Gandayh, H.M. Future of solar energy in Saudi Arabia. *J. King Saud Univ. Eng. Sci.* **2015**, *27*, 153–157. [CrossRef]
3. Fleury-Bahi, G.; Préau, M.; Annabi-Attia, T.; Marcouyeux, A.; Wittenberg, I. Perceived health and quality of life: The effect of exposure to atmospheric pollution. *J. Risk Res.* **2015**, *18*, 127–138. [CrossRef]
4. Labuschagne, C.; Brent, A.C. Sustainable project life cycle management: The need to integrate life cycles in the manufacturing sector. *Int. J. Proj. Manag.* **2005**, *23*, 159–168. [CrossRef]
5. Amiri, M.P. Project selection for oil-fields development by using the AHP and fuzzy TOPSIS methods. *Expert Syst. Appl.* **2010**, *37*, 6218–6224. [CrossRef]
6. Enea, M.; Piazza, T. Project selection by constrained fuzzy AHP. *Fuzzy Optim. Decis. Mak.* **2004**, *3*, 39–62. [CrossRef]
7. Lespier, L.P.; Long, S.; Shoberg, T.; Corns, S. A model for the evaluation of environmental impact indicators for a sustainable maritime transportation systems. *Front. Eng. Manag.* **2019**, *6*, 368–383. [CrossRef]
8. Ligus, M. Evaluation of economic, social and environmental effects of low-emission energy technologies development in Poland: A multi-criteria analysis with application of a fuzzy analytic hierarchy process (FAHP). *Energies* **2017**, *10*, 1550. [CrossRef]
9. Li, L.; Fan, F.; Ma, L.; Tang, Z. Energy utilization evaluation of carbon performance in public projects by FAHP and cloud model. *Sustainability* **2016**, *8*, 630. [CrossRef]
10. Alyamani, R.; Long, S.; Nurunnabi, M. Exploring the Relationship between Sustainable Projects and Institutional Isomorphisms: A Project Typology. *Sustainability* **2020**, *12*, 3668. [CrossRef]
11. Kubler, S.; Robert, J.; Derigent, W.; Voisin, A.; Le Traon, Y. A state-of the-art survey & testbed of fuzzy AHP (FAHP) applications. *Expert Syst. Appl.* **2016**, *65*, 398–422.
12. Bilgen, B.; Şen, M. Project selection through fuzzy analytic hierarchy process and a case study on Six Sigma implementation in an automotive industry. *Prod. Plan. Control* **2012**, *23*, 2–25. [CrossRef]
13. Nguyen, L.D.; Tran, D.Q. *FAHP-Based Decision Making Framework for Construction Projects. Fuzzy Analytic Hierarchy Process*; CRC Press: Boca Raton, FL, USA, 2017; p. 327.
14. Chu, P.-Y.V.; Hsu, Y.-L.; Fehling, M. A decision support system for project portfolio selection. *Comput. Ind.* **1996**, *32*, 141–149. [CrossRef]
15. Huang, C.-C.; Chu, P.-Y.; Chiang, Y.-H. A fuzzy AHP application in government-sponsored R&D project selection. *Omega* **2008**, *36*, 1038–1052.
16. Hatefi, S.M.; Tamošaitienė, J. Construction projects assessment based on the sustainable development criteria by an integrated fuzzy AHP and improved GRA model. *Sustainability* **2018**, *10*, 991. [CrossRef]
17. Sabaghi, M.; Mascle, C.; Baptiste, P.; Rostamzadeh, R. Sustainability assessment using fuzzy-inference technique (SAFT): A methodology toward green products. *Expert Syst. Appl.* **2016**, *56*, 69–79. [CrossRef]
18. Durairaj, S.; Sathiya Sekar, K.; Ilangkumaran, M.; RamManohar, M.; Thyalan, B.; Yuvaraj, E.; Ramesh, S. Multi-Criteria Decision Model for Biodiesel Selection in an Electrical Power Generator Based on Fahp-Gra-Topsis. *Int. J. Res. Eng. Technol.* **2014**, *3*, 226–233.
19. Seddiki, M.; Bennadji, A. Multi-criteria evaluation of renewable energy alternatives for electricity generation in a residential building. *Renew. Sustain. Energy Rev.* **2019**, *110*, 101–117. [CrossRef]
20. Malik, A.; Al Badi, M.; Al Kahali, A.; Al Nabhani, Y.; Al Bahri, A.; Al Barhi, H. Evaluation of renewable energy projects using multi-criteria approach. In Proceedings of the IEEE Global Humanitarian Technology Conference (GHTC 2014), San Jose, CA, USA, 10–13 October 2014.
21. Alyamani, R.; Long, S. Integrating Sustainable Project Typology and Isomorphic Influences: An Integrated Literature Review. In Proceedings of the International Annual Conference of the American Society for Engineering Management, Coeur d'Alene, ID, USA, 17–20 October 2018; pp. 1–10.
22. Saaty, T.L. *The Analytic Hierarchy Process, Planning, Priority Setting, Resource Allocation*; McGraw-Hill: New York, NY, USA, 1980.
23. Kaur, P.; Chakrabortyb, S. A new approach to vendor selection problem with impact factor as an indirect measure of quality. *J. Mod. Math. Stat.* **2007**, *1*, 8–14.

24. Shukla, R.K.; Garg, D.; Agarwal, A. An integrated approach of Fuzzy AHP and Fuzzy TOPSIS in modeling supply chain coordination. *Prod. Manuf. Res.* **2014**, *2*, 415–437. [CrossRef]
25. Ballı, S.; Korukoğlu, S. Operating system selection using fuzzy AHP and TOPSIS methods. *Math. Comput. Appl.* **2009**, *14*, 119–130. [CrossRef]
26. Hsieh, T.-Y.; Lu, S.-T.; Tzeng, G.-H. Fuzzy MCDM approach for planning and design tenders selection in public office buildings. *Int. J. Proj. Manag.* **2004**, *22*, 573–584. [CrossRef]
27. Sun, C.-C. A performance evaluation model by integrating fuzzy AHP and fuzzy TOPSIS methods. *Expert Syst. Appl.* **2010**, *37*, 7745–7754. [CrossRef]
28. Wang, B.; Song, J.; Ren, J.; Li, K.; Duan, H. Selecting sustainable energy conversion technologies for agricultural residues: A fuzzy AHP-VIKOR based prioritization from life cycle perspective. *Resour. Conserv. Recycl.* **2019**, *142*, 78–87. [CrossRef]
29. Shenhar, A.J.; Dvir, D. Toward a typological theory of project management. *Res. Policy* **1996**, *25*, 607–632. [CrossRef]
30. Clarke, H. Evaluating infrastructure projects under risk and uncertainty: A checklist of issues. *Aust. Econ. Rev.* **2014**, *47*, 147–156. [CrossRef]
31. Toma, S.-V.; Chiriţă, M.; Şarpe, D. Risk and uncertainty. *Procedia Econ. Financ.* **2012**, *3*, 975–980. [CrossRef]
32. Stock, G.N.; Tatikonda, M.V. A typology of project-level technology transfer processes. *J. Oper. Manag.* **2000**, *18*, 719–737. [CrossRef]
33. Chen, H.H.; Kang, H.-Y.; Lee, A.H. Strategic selection of suitable projects for hybrid solar-wind power generation systems. *Renew. Sustain. Energy Rev.* **2010**, *14*, 413–421. [CrossRef]
34. Işik, Z.; Aladağ, H. A fuzzy AHP model to assess sustainable performance of the construction industry from urban regeneration perspective. *J. Civ. Eng. Manag.* **2017**, *23*, 499–509. [CrossRef]
35. Luthra, S.; Kumar, S.; Garg, D.; Haleem, A. Barriers to renewable/sustainable energy technologies adoption: Indian perspective. *Renew. Sustain. Energy Rev.* **2015**, *41*, 762–776. [CrossRef]
36. Solangi, Y.A.; Tan, Q.; Mirjat, N.H.; Valasai, G.D.; Khan, M.W.A.; Ikram, M. An integrated Delphi-AHP and fuzzy TOPSIS approach toward ranking and selection of renewable energy resources in Pakistan. *Processes* **2019**, *7*, 118. [CrossRef]
37. Buckley, J.J. Fuzzy hierarchical analysis. *Fuzzy Sets Syst.* **1985**, *17*, 233–247. [CrossRef]
38. Chang, D.-Y. Applications of the extent analysis method on fuzzy AHP. *Eur. J. Oper. Res.* **1996**, *95*, 649–655. [CrossRef]
39. Kahraman, C.; Cebeci, U.; Ruan, D. Multi-attribute comparison of catering service companies using fuzzy AHP: The case of Turkey. *Int. J. Prod. Econ.* **2004**, *87*, 171–184. [CrossRef]
40. Švajlenka, J.; Kozlovská, M. Evaluation of the efficiency and sustainability of timber-based construction. *J. Clean. Prod.* **2020**, *259*, 120835. [CrossRef]
41. Kahraman, C.; Kaya, İ.; Cebi, S. A comparative analysis for multiattribute selection among renewable energy alternatives using fuzzy axiomatic design and fuzzy analytic hierarchy process. *Energy* **2009**, *34*, 1603–1616. [CrossRef]
42. Amer, M.; Daim, T.U. Selection of renewable energy technologies for a developing county: A case of Pakistan. *Energy Sustain. Dev.* **2011**, *15*, 420–435. [CrossRef]
43. Švajlenka, J.; Kozlovská, M.; Pošiváková, T. Analysis of selected building constructions used in industrial construction in terms of sustainability benefits. *Sustainability* **2018**, *10*, 4394. [CrossRef]
44. Fu, H.-H.; Chen, Y.-Y.; Wang, G.-J. Using a Fuzzy Analytic Hierarchy Process to Formulate an Effectual Tea Assessment System. *Sustainability* **2020**, *12*, 6131. [CrossRef]
45. Tsai, H.-C.; Lee, A.-S.; Lee, H.-N.; Chen, C.-N.; Liu, Y.-C. An Application of the Fuzzy Delphi Method and Fuzzy AHP on the Discussion of Training Indicators for the Regional Competition, Taiwan National Skills Competition, in the Trade of Joinery. *Sustainability* **2020**, *12*, 4290. [CrossRef]

Article

Analysis of the Impact of COVID-19 on the Food and Beverages Manufacturing Sector

Arnesh Telukdarie [1,*], Megashnee Munsamy [1] and Popopo Mohlala [2]

1 Postgraduate School of Engineering Management, University of Johannesburg, Johannesburg 2006, Gauteng, South Africa; mmunsamy@uj.ac.za

2 Food and Beverage Manufacturing Sector Education and Training Authority, Johannesburg 2006, Gauteng, South Africa; PopopoM@FoodBev.co.za

* Correspondence: arnesht@uj.ac.za; Tel.: +27-61-450-5948

Received: 3 September 2020; Accepted: 20 October 2020; Published: 10 November 2020

Abstract: The globe has been subjected to an unprecedented health challenge in the form of COVID-19, indiscriminately impacting the global economy, global supply chains, and nations. The resolution of this unprecedented challenge does not seem to be in the short-term horizon but rather something the globe has to live with. Initial data provides for some insights on responses, precautions, and sustainability protocols and processes. The Food and Beverages Manufacturing sector in South Africa (SA) and globally is an expeditious respondent to the COVID-19 challenge. Food is essential for human existence, but the food value chain is subjected to significant COVID-19 risks. The Food and Beverage Sector Education and Training Authority is responsible for skills development in the Food and Beverages (FoodBev) Manufacturing Sector in South Africa and seeks to quantify Foodbev sustainability. This research paper reviews global literature, performs a high-level knowledge classification, with the aim of expedited awareness, knowledge sharing, and most importantly, quantification of an expedited response, within the FoodBev Manufacturing sector in SA. The research is contextualized via a SA sector-based instrument deployment and data analysis. The paper provides insights into COVID-19 impact, adaptations, and responses in the SA Food and Beverages Manufacturing sector.

Keywords: COVID-19; food and beverage; manufacturing; sustainability

---

1. Introduction

In the context of the global economy, food is key to consumer confidence, global peace, and personal sustenance. Maintaining the movement of food through the global supply chain is essential for sustaining life. There is an overwhelming global response to the COVID-19 virus with Food and Beverage Sector sustainability a key challenge. As the Food and Beverages Manufacturing sector provides the essential service of food, a significant portion of businesses remained open in South Africa (SA) during the lockdown period with the exception of the alcohol manufacturers and distributors. However, there are disruptions to the global and local supply chain impacting the production of food. The COVID-19 response time and the knowledge base for preparing the sector in SA is fundamental.

The COVID-19 challenge faced by the Food and Beverages Manufacturing sector includes operations, safety, supply chain, training, emergency responses, awareness, incident management, recreating business models, digitalization, and other unanticipated impacts. Further to this, COVID-19 has changed consumer behavior to food. There are significant global and South Africa (SA)-specific impacts anticipated, with Small, Medium, and Micro Enterprises (SMME)'s expected to be significantly impacted. Thus, the short term and long-term impacts of COVID-19 on the South African Food and

Beverages Manufacturing sector has to be determined to expedite recovery and to develop measures for readiness should another such disruption occur.

The SARS COVID-19 is categorized as a global pandemic [1], with almost 865,000 global deaths and 26 million infections, as of 3 September 2020 [2]. COVID-19 transmission occurs via contact and proximity, is carried on surfaces for up to nine days, and is destroyed via limited protocols such as alcohol, UV light, and 0.5% sodium hypochlorite [3].

The Food and Beverages Manufacturing sector is in urgent need of an understanding the virus, from an operations perspective. The Food and Beverages Manufacturing sector has to institute health and safety protocols to respond to employee safety. Key additional considerations for the sector relates to operational constraints. The need to restructure operations so as to produce the relevant food products in a safe, financial, and environmentally sustainable manner.

The research team seeks to also identify catapult opportunities via this study. This research seeks to address the following questions:

- What is the global best practice in planning and responding to pandemics such as COVID-19?
- What are the impacts, both short term and long term, on the South African Food and Beverages Manufacturing sector?
- What are the recovery/mitigation measures required?

The high-level methodology of this study is mixed methods and includes a global peer reviewed literature search, a white paper search, compilation of benefit determination in SA, and an instrument to evaluate relevance and benefit in the SA Food and Beverages Manufacturing context. This research commences with a global scan of current knowledge, albeit not limited to peer reviewed publications. The objective is to rapidly gather relevant contextualized information. The research team build a South African (SA) data collection and response protocol based on this initial data collection cycle. The key knowledge themes are developed as the qualitative data is collected.

1.1. *Impact of COVID-19 on the Food and Beverages Manufacturing Sector Globally and in South Africa*

The research team adopt a multi-tier literature review approach that includes peer reviewed journals and conference papers together with a review of white papers from globally recognized entities. A Scopus search is conducted to commence publications analysis. Figure 1 illustrates the countries with the highest number of peer-reviewed COVID-19 publications, specific to the Food and Beverages Manufacturing sector. A total of 482 publications are identified, screened, and reviewed as a basis for this paper. In addition, a global best practice search is also conducted in order to find further papers on Food and Beverages Manufacturing and COVID-19.

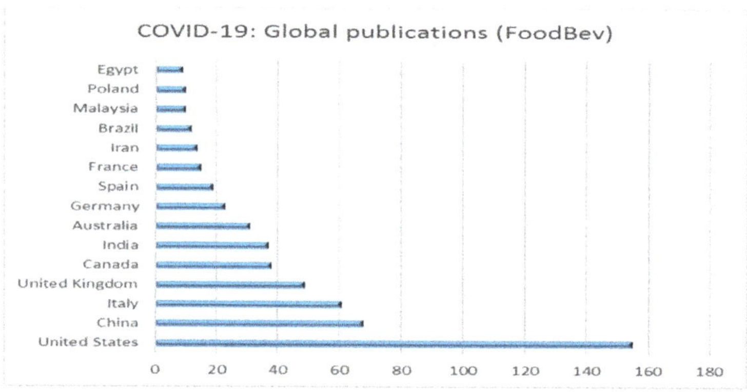

Figure 1. Number of publications per country.

It is projected that the South African food and non-alcoholic beverages manufacturing is going through a moderate decline of flow of income (−10% to −30%), whereas the alcoholic beverages will experience a bigger impact at −60% [4]. Further, Anderson et al. [5] argue the potential for significant interruption in the activities of the Food and Beverages Manufacturing sector cannot be ignored.

The varying levels of lockdown initiated in various countries globally, social distancing requirements and health concerns, have seen a shift in consumer behavior to home cooked meals rather than eating out [6]. Beef, which is used extensively in restaurants and fast food services, has experienced a significant decline in demand. Since the start of lockdown, an increased demand of snacks and baking goods was observed globally, as people were eating more during the day [7]. In Italy and China, changes in food patterns were observed with greater demand for staple, non-perishable foods and a reduction in demand for fresh fruit and demand [8]. This could lead to wastage due to limited shelf life and loss of income to farmers. Further, the high value commodities processing plants are labor intensive and maintaining of social distancing is difficult. The closure of restaurants, coffees shops, and farm-holiday stays resulted in a 10% reduction in milk consumption, a significant impact to the dairy industry [8]. In the international sphere the confectionary, snacks, other food products, baking, and grain mill industries have experienced an increase in demand of products and have been forced to operate on online platforms to meet the demand. However, the baking industry might experience challenges as lockdown continues. The executive director for the SA Baking Chamber reported that Tiger Brands closed one of their biggest firms during lockdown, but indicated that the supply of food remains stable. The company's profits have fallen by 75% (R359.6 million in 2020 vs. R1.4 billion in 2019) during the same time last year due to COVID-19 [9]. Demand for baked food could prove a challenge for food producers, especially producers of soft luxuries, as consumption is a function of income as well as certainty of income. The latter is a result of other people's financial status as they have lost their jobs and therefore do not have spending power. South Africa's alcoholic beverage industry has been hit hard by COVID-19 lockdown regulations, as the sale and export of alcohol is prohibited under the disaster regulations governing the national lockdown in South Africa [10]. Reports show that an estimated 117,600 jobs have been lost throughout the alcohol industry, 13% of the craft beer sector is in the process of shutting up shop, and the wine industry is in severe distress [11].

The suppliers of the Food and Beverages Manufacturers have also been impacted by COVID-19, influencing manufacturing operations. High value commodities of fresh fruit, vegetable, and fish are affected by the potential shortage of migrant seasonal labor for fruit picking, harvesting, etc. [8,12]. According to FAO [13], the fishing industry in many countries is considered critical for food production and supply; however, due to COVID-19, the fishing industry is facing two main problems of disrupted supply chain and declining markets, resulting in disruptions in transportation, trade, and labor. Disruptions in these lead to delayed stock and lower supplies, access, and consumption of these food [13]. Fish farmers who cannot sell the fish must now keep larger volumes of fish that need to be fed, increasing costs and risks [14]. The production of aquaculture may also be affected by the ability to purchase feed and seed due to cargo restrictions and prioritization [14]. Health safety measures of physical distancing and face masks can make fishing difficulty even to the point of reducing or stopping it [14]. If a crew member has the virus, it can easily spread to other members, and medical assistance may not be easily accessible due to being out at sea [13]. The reduced staff may result in a decline of Fisheries Monitoring and Control as observed during the 2013–2016 Ebola outbreak in West Africa [14]. This results in an increase in illicit activities and in a less responsible level of managing, monitoring, and controlling of fishing operations. The staple commodities of wheat, maize, corn, and soybean are capital intensive farming and could be affected by value chain disruptions due the requirements of various and large amounts of inputs such as seeds, fertilizers, pesticides, diesel, etc. [15]. This does not have a short-term impact but a long-term impact of food growth and availability for the following seasons/years. Restriction of movement and closure of markets have a significant impact on small holder producers, who do not have the finance and resources of larger producers [12]. A prolonged disruption could affect the ability to resume production [8].

1.2. COVID-19 and Sustainability

The lockdown enforced in various countries globally and the operational impacts of COVID-19 have had negative impacts on economies and severely impacted the sustainability of businesses. To stimulate economic recovery and towards business continuity, countries have initiated various measures, see Table 1, from financial stimulus packages to delay in loan repayments to waiver of loan fees to technology support. The measures introduced by a few countries are detailed below.

Table 1. Country specific COVID-19 responses.

Country	Response
China [15]	Alibaba, an e-commerce company, created an e-commerce platform to assist famers in selling their unsold agricultural products and is also creating a "green channel" for fresh produce. The Beijing central government set up a USD 20 million subsidy for purchasing of machines and tools for agricultural purposes. It is also providing low interest rate loans and rent reductions to incentivize development of digital agricultural technologies, such as agricultural drones and unmanned vehicles. The Agricultural Bank of China is reducing the interest rate by 0.5% for small and medium enterprises, self-employed, and private owners in the Hubei Province [8].
Italy [15]	Allocated 100 million euros in support of agricultural and fishing businesses, which had to halt operations. A further 100 million Euros is allocated for financing. Permitting advance payments from the European Union subsidies to farmers.
India [16]	Implementing software for warehouse-based trading of harvested food to reduce congestion at wholesale markets.
Poland [16]	Offered subsidized loans to food processing businesses to facilitate continuous operations.

The diversification of business offerings from products and services to business supply chain influences the business ability to respond to business disruptions and uncertainties. Alcohol producers in the UK such as William Grant & Sons and Diageo have begun producing hand sanitizers, whilst chemical company INEOS set up a hand sanitizer facility manufacturing one million bottles per month, which began production on 31 March 2020 [17]. A ventilator consortium was set up in the UK to address the need of increased demand. The consortium consisting of various businesses such as Airbus, BAE System, Microsoft, Accenture, Ford Motor Company, and Rolls-Royce announced on 2 April 2020 they are to produce 1500 ventilators per week based on the Penlon and Smiths design. The consortium is reported to have received an order for over 15,000 units from the UK government alone [17]. These businesses may not continue manufacturing the products post COVID-19, but it allowed the business to be operational and productive during a disruptive period.

A key consideration in business sustainability is environmental sustainability. The environmental impact of COVID-19 globally is initially quantified to be positive [18] due to reduced traffic, movement, air travel, and energy demand. Rugani and Caro [19] determined a 20% reduction in carbon footprint in Italy for the months of March and April 2020 in comparison to the same months for the 2015–2019 period, an actual reduction between 5.6 and 10.6 Mt CO_2e.

However, there are concerns the post COVID-19 environmental impacts might be neglected in the drive to stimulate economic resurgence. Helm [18] states COVID-19 responses may result in significant post COVID-19 growth which could have a negative impact on climate change. Rosenbloom and Markard [20] state that the US may be revitalizing the fossil fuel industry via stimulus funding, whilst a report by the German Council of Economic Experts neglected to mention climate change or sustainability. However, 17 European climate and environment minsters requested a Green Deal be incorporated in the recovery plan. Rosenbloom and Markard [20] recommend the recovery plan include funds to support low carbon energy technologies/projects that reduce or mitigate carbon emissions such as renewable technologies, energy storage, and electric vehicle. They further recommend support and development of remote working, video conferencing, e-commerce, and reduced travel, whilst suggesting using the current destabilization of business and economies to accelerate transition towards clean options. Other considerations for climate change include the impact of elevated alcohol manufacture and application as a preventative measure as well as the impact of COVID-19 waste on the environment.

1.3. COVID-19 Impact on Supply Chain

COVID-19 has impacted all spheres of the global supply chain including distribution and packaging, as well as sourcing of raw materials [21,22]. Lockdowns disrupted the transportation of packaged foods, prepared foods, and non-alcoholic and alcoholic beverages [21], whilst some companies had to close for up to two weeks for cleaning purposes [15,23]. Weersink et al. [23] also identifies capacity constrains due to social distancing in the workplace leading to operational challenges. Operational challenges include the nature of packaging due to the reduced restaurant demand.

The disruption of the global supply chain has emphasized the risk of high probability external dependence on essential items. The economic stability of a country impacts its food security; the ability to manufacture or import the required food needs. Singapore, which imports a large amount of its food, plans to produce 30% of its food by 2030, in comparison to 2019 where it only produced 10% of its food needs [24]. Hossain [25] highlights various eastern block countries limitation of imports and exports on food products, including rice, poultry, eggs, vegetables, fish, and fruit. Gemmill-Herren [26] further identifies countries would procure food locally for various reasons, beyond the fact that imports have slowed down. Thailand, which depends heavily on tourism, approximately 20% of GDP, plans to diversify its economic sector through target industries of electronics, automotive, and chemicals, as its manufacturing sector already contributes 27% of its GDP [24].

Weersink et al. [23] also highlights central co-ordination bodies for farmers and distribution to be a priority. In Brazil, the existing Solidarity Purchasing Groups (GAS) and Community Supported Agriculture (CSA) exist as a network for farmers. This is an example of localized supply chain management, should localized networks be prioritized. Other strong examples of localized community coalitions are presented in North America and other parts of the world. Preiss [27] also identifies local supply networks.

The literature also indicates that a broader system of systems or ecosystem approach should be considered for workforce and supply chain; workforce ecosystems should include health and childcare as priorities while supply chains need to become dynamic and localized. The need to adopt a distributed global services model is recommended.

- Use a mixture of service models to de-risk the organization in a volatile world. Distributed global services mean that high performance can be delivered anytime, anywhere.
- Improvement of call centers and customer interaction.
- Augment and automate service: Virtual agents.
- Enhance virtual agent capabilities to support COVID-19 specific requests or growing Business as Usual volume.
- Alternative suppliers and distributors.
- Visibility of supply and demand network (digital).

1.4. COVID-19 and Technology

COVID-19 is forcing the development of resilient food systems [28]. Digital technologies can improve resiliency of food chains and assist in optimizing outputs [22,29]; 4IR technologies such as Big Data, Internet of Things (IoT), Cloud Computing, Robotics, and Automation facilitate remote and autonomous working whilst providing transparency in operations. Galanakis [30] reinforces various aspects of digital and 4IR, specific to benefits to the sector including supply chain optimization, and faster time to market. Preiss [27] identifies e-commerce platforms as a key to surviving COVID-19. These platforms are an opportunity in Brazil. Weersink et al. [23] identifies the process of automation to be accelerated at all levels of the supply chain in the Canadian food sector. Contactless delivery is becoming a norm from farm to store [23], requiring digital technologies.

Further consumers require more information about the products being purchased. Consumers are digitally savvy and smart devices deliver knowledge, provided for automation (personal data, searchers for choices, order management). Barcodes with embedded microscopic electronic devices such as Radio Frequency Identification (RFID) tags, genetic markers, and hyperspectral imaging used in conjunction with mobile phones would enable consumers to access information about the authenticity, freshness, ripeness, shelf life, and nutritional content of food [31]. Intelligent packaging is driving the growth of the packaging industry because it integrates an intelligent (communication) aspect to conventional packaging as it communicates information to the consumer in real-time as and when it senses, detects, or records any changes to the product [32]. Consumers are also seeking to adopt technology in advising on food to eat. This is specific to technologically centric personal analysis.

With social distancing requirements, remote working has increased, via the use of various platforms such as Zoom and Microsoft Teams, which further enhances resiliency; business can contact suppliers and customers and develop alternative collaborative solutions.

1.5. Agile Workforce

The COVID-19 pandemic has highlighted the need for a responsiveness workforce inclusive of management. It is clear that food and beverage companies must adopt to operate in a volatile, uncertain, and complex environment. Companies have to prepare for changes in operations; this includes a structured, specialized management response team as recommended by Accenture [33] (refer to Figure 2).

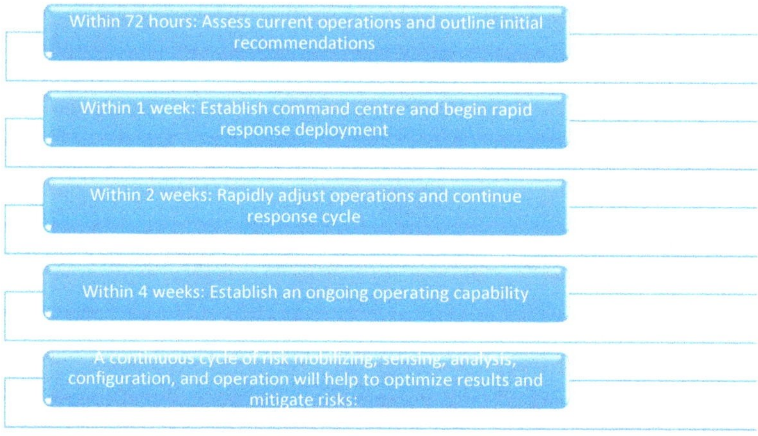

Figure 2. Management response timeframe (Adopted from Accenture [33].)

COVID-19 has also emphasized the need for an agile and elastic workforce [34]. Accenture [33] recommends the creation of an elastic digital workspace taskforce comprising of HR, legal, business, IT, marketing, sales, and communication representatives to oversee activities; a central command center. One of the most adopted techniques of an agile workforce is remote working, facilitated via collaboration platforms such as Zoom or Microsoft Teams. Accenture has identified six aspects requiring address to transition to a remote workspace [34]:

Culture and adoption: It is critical that the necessary tools, training, and coaching are provided. The use of online team meetings or videos: to keep staff aware of changes and the need for the change and demonstrations/trainings sessions on the use of the digital tools. The transition to cloud: for use of applications/software, creating and collaborating on documents and task management. It is also recommended to develop recommended/best practices for remote working, such as turning off video to improve audio in team meetings or workshops.

- Elastic collaboration: Expansion of communication and collaboration tools is required as more workers are remotely based. Ensure employees can use the designated communication and collaboration tools.
- Virtual work environment: Ensure employees have the tools and access to relevant applications and most especially data.
- Seamless networking: Ensure seamless connection to required business networks, cloud applications/services, and partner networks.
- Distributed continuity: Pandemic preparation must be merged into business plans. Keeping well-informed of developments and availability to make quick decisions and communication.
- Adaptive security: The mandatory security procedures are in place to decrease corporate risk and safety breaches. Develop and increase the usage of analytics and automation to reduce the amount of human intervention required.

The key is to develop a team that can respond quickly in an emergency situation [12]. This may form part of business continuity planning or can become a system centric approach. Other considerations for this strategy include the adoption of Artificial Intelligence (AI) [33], where the system is empowered via data and history to respond to the "threat."

Furthermore, in developing an agile and responsive workforce, food safety has to be inherent and inclusive of supply chains [35]. Food Safety Management Systems (FSMS) based on the Hazard Analysis and Critical Control Point (HACCP) principles must be put in place to manage food safety risks, and prevent and respond to food safety emergencies like outbreaks of foodborne diseases. Dyal et al. [36] highlights disinfecting of high touch areas, introducing outside break areas and adjusting start/stop and shift times. In developing the agile workforce and workspace, the following must be considered towards ensuring food safety:

- Reconfiguration of workplace: The workplace has to provide for social distancing. The number of people on site and within a defined confined space has to be reviewed.
- Worktime re-adjustment: Change in schedules, introducing shifts and other changes to worktimes has to be considered.
- Hazardous pay: The prioritization of categories of work based on risk has to be reviewed. Certain categories of workers would qualify for hazardous pay.
- Making remote working permanent: Staff that can work offsite should be contracted to do so. This should be encouraged with productivity measures in place. The impact on company real estate, onsite facilities, travel, and the environment should be additional considerations.

The literature also discusses the option of companies considering labor exchange. In principal, companies create a pool of labor that can be activated based on various business constrains. The key is to have a skilled labor force that can be interchangeably used based on worker health, demand, and other factors.

Additional Human Resources (HR) aspects have to change [36]. This includes the approach to sick leave. Rewarding sick leave would drive the wrong behavior. Additional sick leave, liability benefit which is not punitive.

1.6. COVID-19 and Skills

The Food and Beverages Manufacturing SETA's core mandate is skills development specific to the Food and Beverages Manufacturing sector. This research focuses on COVID-19 mitigation specifically via skills development for sustainability. The shape of skills training has changed globally especially with the onset of COVID-19. Various training mechanisms have to be addressed.

The introduction of asynchronous (students learn the same content at different times) learning is a consideration, as not all learners need to be in the same classroom at the same time. Digital content distribution makes asynchronous learning a reality [37].

Latchem [38] highlights various tools that can be successfully used for skills development for post school learning; the book focuses on Technical and Vocational Education and Training (TVET). The following methods are discussed in detail: Virtual Reality; Mobile Learning; Simulations, Games and Role Plays; Augmented Reality; 3D Printing; Digital Repositories. The book provides insights into the work done in various countries globally. The use of Information and Communication Technologies (ICT) is highlighted as a driver to improve throughput with cost reductions and improved access. Latchem [38] also highlights the need to understand that a different approach is required for ICT-based learning, including:

- Discipline
- Pre-screening of students and basic skills development
- Hardware and data
- A structured mix of theory and practical is required
- Need to further train the SMME sector on the skills of ICT instructions

Latchem [38] identifies mixed time investment for learners as working well; this includes a 40% allocation by the companies employing the learners and 60% of learner's own time. Timely response mechanisms must be put in place, including verbal communication with learners. Verawardina [39] reinforces the need for structure in the learning cycle with roles and responsibilities as an additional highlight.

2. Materials and Methods

The aim of this study is to extract a SA specific response to COVID-19, based on current international trends. The research methodology adopted for this study is best described as mixed methods, comprising of qualitative data collection via a desktop literature review and a quantitative instrument deployment to gather relevant data, specific to the South African Food and Beverages Manufacturing context.

The initial aspects of the study include a global literature search, seeking response and impacts specific to the Food and Beverage manufacturing sector, and COVID-19. Based on the literature analysis, a structured data gathering instrument is developed and deployed in the SA Food and Beverages Manufacturing Sector. The instrument aligns to literature as extracted in the literature section.

Sampling considerations include the fact that the food and beverage sector in SA comprises some 11,000 companies, of which 93% are SMME's and around 700 companies provide Work Skills Plans (WSP) to the Food and Manufacturing Sector Authority (SETA). The SETA databases are used to manage the data gathering processes. A statistically representative sample for the 11,000 members would be 184, for a 95% confidence and 1% error.

The instrument is developed and with the knowledge of the literature categorized into sections, see Figure 3.

A thematic analysis is conducted based on the sector, company size, and province in which the company operates its business. The Cronbach Alpha reliability test is conducted on feedback data for each knowledge theme. The research team conduct thematic analysis on size of company, supply chain impacts, skills impact, and operational impacts and resources centric impacts. Long-term COVID-19 impact and sustainable responses to COVID-19 are extracted.

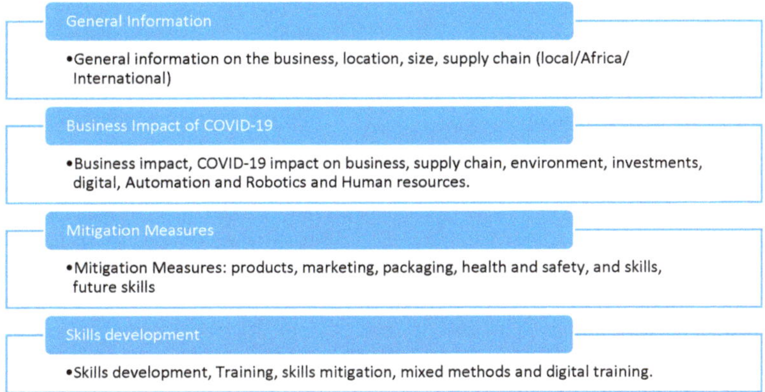

Figure 3. Instrument categories and focus.

3. Results

3.1. Statistical Analysis of Responses

The overall validity of the study is tested based on the population and number of responses received. The team received 106 responses, which gives a 90% confidence and an 8% error. The first tier of analysis conducted by the team relates to the data reliability tests. The Cronbach Alpha statistical test is adopted to test for reliability. Due to diversity of subject matter of questions, within the COVID-19 context, the questions are clustered bases on themes such as human resources, supply chain, training questions. The instrument is tested for reliability in sections and as per the instrument design. The following reliability tests are conducted with Cronbach Alpha results detailed in Table 2.

Table 2. Cronbach Alpha results.

No.	Question Description	No. of Sub-Categories	Cronbach Alpha
Q1	At this point in time, how is your business operation being impacted by COVID-19 and what is the level of impact?	11	0.76
Q5	Studies show that because of the unprecedented measures being taken globally to attempt to limit the spread of the virus, there have been significant disruptions in food supply chains. What kind of impact is COVID-19 having on supply chains of your business operation?	6	0.81
Q8	Q8: Is your company using or planning to use any of the following online and/offline distance learning tools for training: Online learning programs, video conferencing, virtual reality simulators, and multimedia.	5	0.586
Q10	How is your company handling HR issues as they relate to COVID-19?	5	0.58
Q15	Businesses across the world have put in place a variety of measures to help mitigate against the impacts on their business. How has your company responded to the COVID-19 outbreak?	17	0.71

The first tier of analysis conducted by the team relates to the data reliability. The Cronbach Alpha statistical test is adopted to test for reliability. Due to diversity of subject matter of questions, within the

COVID-19 context, the questions are clustered based on themes such as HR, supply chain, training, impacts, and mitigations. The Instrument is tested for reliability in each of the clusters.

Questions 1, 5, 10, and 15 are defined as multi-category questions, as each question comprises an overarching question, with the response delineated to applicable themes. In Question 1, the companies rate the impact of operational challenges introduced by COVID-19. These challenges are identified from a global review of business challenges brought upon by COVID-19. A Cronbach Alpha of 0.76 is calculated indicating data reliability and that South African Food and Beverages Manufacturing companies (large, medium, micro, and small) are encountering similar operational challenges. It further indicates that SA Food and Beverages Manufacturing companies are experiencing similar challenges as their global counterparts.

Question 5 analyzes the impact to business supply chain, with the Cronbach Alpha calculated as 0.81, indicating a strong reliability. This demonstrates that most companies are experiencing supply chain disruptions, both locally and internationally. This is one of the most significant challenges of businesses globally, given the closure of borders and businesses (due to lock downs).

Question 10 analyzes the business response to HR practices ranging from sick leave policies to COVID-19 response of screening, Personal Protective Equipment (PPE), etc. The Cronbach Alpha of 0.58 indicates some inconsistency. This inconsistency is further analyzed in this document.

Question 15 analyzes the response to 17 mitigation measures identified from a global review. A Cronbach Alpha of 0.71 is calculated indicating data reliability. This demonstrates that SA Food and BEV companies have already or are considering the mitigation measures, aligning to global trends in business recovery.

Questions 2 and 3 address climate concerns, thus they are analyzed together. A Cronbach Alpha of 0.51 is calculated indicating some inconsistency. This inconsistency is due to the disassociation between the questions. Most companies indicated no investment plans aligned to stimulus packages (Q2), but indicated future investment plans in environmental footprint reduction (Q3). The stimulus packages may present investment opportunities in new technologies and processes, which could simultaneously improve business performance and resiliency, while reducing environmental footprint.

Question 4 and 13 are clustered as it evaluates COVID-19 impact on business continuity. A Cronbach Alpha of 0.61 is calculated, indicating some data reliability. Whilst 86% of companies did not envisage closure of the business, 32% identified the COVD-19 impact as severe. This contrasting data could attribute to the low Cronbach Alpha.

Questions 7 and 8 are clustered as it evaluates the current status of adoption of mixed methods in skills development/training. A Cronbach Alpha of 0.59 is calculated, indicating some inconsistency. Sixty-one percent of companies responded to having already adopted some form of mixed methods in their training programs (Q7). However, in evaluation of the specific mixed methods adopted (Q8), inconsistency is observed; 61% state have already adopted video conferencing, 27% each have adopted online learning and multimedia, and 12% have already adopted virtual reality simulator. The inconsistency in the Cronbach Alpha could potentially be attributed to the low adoption levels of online learning, multimedia, and virtual reality simulators.

3.2. Data Analysis

The research team provides an analysis of the overall response prior to sectional, theme, and individual questions analysis. The FB sector covers all nine provinces in SA, which comprises Small, Medium, and Micro enterprises (SMME) and comprises various sectors. An initial analysis is conducted, see Table 3.

The company distribution provides for data on representation and where necessary is used in the sectional, themed, and individual analysis.

Table 3. Provincial/Category/Company size classification.

Row Labels	Companies	Operating Provinces	No. of Companies	Business Sector	No. of Companies
Large	28	Northern Cape	8	Fish	8
Medium	41	Free State	16	Fruits & vegetables	17
Micro	9	Eastern Cape	20	Meat	16
Small	27	Gauteng	45	Baking	16
Grand Total	105	Limpopo	17	Oil & fats	6
		North West	11	Grain Mill	4
		KwaZulu Natal	38	Soft drinks & water	17
		Western Cape	50	Beer & Malt	13
		Mpumalanga	11	Other food products	34
		Meat	16	Dairy	7

3.2.1. Overall Analysis of Impact of COVID-19 on Food and Beverage Manufacturing Sector Businesses

A key screening question reveals that 12 of the 106 companies could close in the next 12 months. Table 4 below provides data relating to size of companies. It is apparent that smaller companies are being impacted with no large companies envisaging closure. The data on category of companies is not clear but of these 12 companies, 6 are from the "meat" category (a total of 16 meat companies responded). The exact driver for such a large representation in the meat category is beyond this study analysis.

Table 4. Business closure timelines.

Closure Timeline	No. of Companies	Small	Medium
1 Month	1		1
3 Months	5	4	1
6 Months	4	2	2
12 Months	2	1	1

The research team sought to reinforce data beyond the current perspective with a question on forecasting. *At present, the long-term impacts of COVID-19 are uncertain, but the ramification will continue to be felt even after the spread of the virus is contained. What will be the impact for the future?*

With 30% of the companies electing not to answer, five percent felt there would be no impact in the future while the remaining 65% of the companies are envisaging future impacts.

The research now focuses on the operational impact of COVID-19 on Food and Beverage Manufacturing companies. Various literature sources provide inputs from a complete shutdown of business, adjusting work hours, impacts on sales and orders, business credit, reduced/increases in selected product lines, employee wellness, and administrative challenges. These are incorporated into a business impact question with a Low/Medium/High impact as feedback from companies. The data (Table 5) clearly indicate that there has been a high impact on small and micro enterprises in all categories. Large and medium companies have experienced low impacts.

Table 5. Analysis of COVID-19 impacts on business operations.

	Large			**Medium**			**Micro**			**Small**		
	Low	Medium	High	Low	Medium	High	Low	Medium	High	Low	Medium	High
We have completely closed our physical place of business	46%	7%	7%	32%	5%	5%	0%	11%	22%	30%	22%	26%
We have adjusted our hours of operation	25%	25%	7%	17%	17%	5%	0%	11%	22%	30%	15%	33%
Temporary shutdown	39%	14%	4%	27%	10%	2%	0%	11%	11%	52%	15%	15%
Orders are being cancelled	43%	14%	0%	29%	10%	0%	0%	0%	33%	37%	11%	30%
Our supply chain is interrupted	25%	25%	7%	24%	10%	32%	0%	0%	33%	30%	19%	30%
We are experiencing decreases in sales	18%	36%	4%	20%	15%	34%	0%	0%	33%	26%	15%	41%
Employee absence due to sickness	29%	25%	7%	32%	22%	15%	22%	0%	11%	41%	33%	4%
Reduced logistics services	18%	39%	0%	20%	24%	22%	0%	0%	33%	30%	37%	15%
The market is causing us to draw our line of credit	39%	7%	11%	22%	12%	29%	0%	0%	22%	30%	19%	30%
Increased administrative bottlenecks	18%	21%	18%	20%	17%	32%	11%	11%	11%	26%	37%	19%
We are experiencing increases in consumer demand for certain products	25%	7%	25%	27%	17%	24%	22%	0%	0%	52%	19%	7%

The data on complete shutdown of business indicates a higher impact on small and micro businesses with large and medium experiencing at least 20% of businesses with a high impact. Essentially 1 in 5 small/micro businesses have been physically closed with a high impact. The adjustment of work hours has had a high impact on small and micro with a low impact on large business. Order cancellations are having a high impact on small business with very little impact on medium to large business. A decrease in sales impacts all but large businesses. Reduced logistics services is a high impact challenge across all sizes of business. Finance and line of credit have been used by medium/small/micro businesses and has a high impact. Small businesses indicate an increase in administrative bottleneck, this cannot be internal. It may be external with unemployment insurance and other government dependent activity.

3.2.2. Supply Chain and FB Manufacturing

The initial analysis is conducted on the supply chain as a key theme; the literature section of this paper provides details on supply chain disruptions.

With the onset of COVID-19, international trends are that supply chains are challenged. An analysis is conducted/size of respondents. With reference to Figure 4, it is very apparent that most companies (in excess of 80%) source locally, with smaller companies more dependent on local supply than larger. On the outbound side, domestic is favored by smaller companies with larger companies conducting significant international trade. The African Region is the smallest focus both on inbound and outbound.

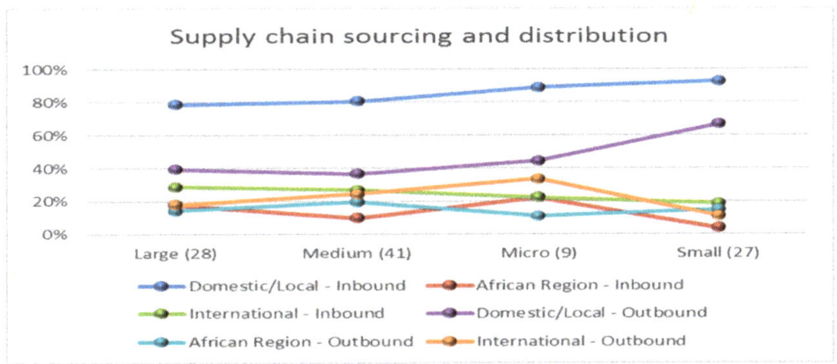

Figure 4. Analysis of impact on inbound and outbound network.

The research team added a question on if companies, due to COVID-19, sought to join SA sourcing and distribution networks and 50% of the current responders indicate that they currently source outside of SA and are looking at transitioning to SA sourcing. *Data on supply chain is further reinforced by Section B/Business Impact. Question 1: What is the level of impact:*

- *Our supply chain is interrupted.*
- *Reduced Logistics services.*

The results indicate that the combined medium to high impact in all sizes of business ranged from 36% to 50%. This indicates the impact of COVID-19 on supply chain and logistics to be a challenge in the SA context.

3.2.3. Business Impact: Question 5

The data (Table 6) indicates clear challenges in lower domestic sales to customers and businesses, with medium to high impact on all business sizes. Accessing materials and goods internationally is also a challenge to business. It is also apparent that small businesses experience the most significant impacts on supply chain and sourcing and distribution. Further data probing on supply chain is conducted by the questions.

Table 6. Analysis of impact on supply chain network.

	Large			Medium			Micro			Small		
	Low	Medium	High	Low	Medium	High	Low	Medium	High	Low	Medium	High
Difficulty accessing materials, goods and services within South Africa	21%	32%	7%	15%	22%	5%	0%	22%	11%	37%	26%	11%
Difficulty accessing materials, goods and services internationally	11%	21%	21%	7%	15%	15%	11%	11%	11%	44%	19%	7%
Difficulty exporting goods and services	14%	18%	25%	10%	12%	17%	11%	11%	11%	33%	15%	22%
Difficulty importing goods from abroad	18%	18%	21%	12%	12%	15%	0%	11%	22%	41%	7%	19%
Lower domestic sales to consumers	14%	36%	7%	10%	34%	24%	0%	11%	22%	22%	19%	37%
Lower domestic sales to business	11%	43%	7%	17%	29%	22%	0%	11%	33%	26%	19%	33%

Regions and countries globally are preparing to improve supply chain via locally co-ordination. Does the company plan to institute or participate in a centrally coordinated (This may be on a provincial/ National level)? Yes/no

Research shows that COVID-19 has put a renewed urgency behind automation and the use of robotics to mitigate against the disruptive impact of supply chain. Is digitalization a consideration in local digital networks? Yes/no

The data, Table 6, indicates that 58% of large companies are considering a centrally coordinated supply chain network, either at a local or national level. Only 27% of medium size and 26% of small companies are considering a centrally coordinated supply chain network. A similar trend is observed in the evaluation of adoption of digital supply chain networks, with medium and small companies demonstrating lower consideration of adoption, whilst large companies (63%) are strongly in support of digital supply networks.

3.2.4. Company Resource Based Mitigation

A key resource at every company, small or large, are the employees. In this period of COVID-19, companies have to take care of this important resource. Best practice literature provides for responses in HR practices, leave, and healthcare, a dedicated COVID-19 HR team, dedicated risk teams, and communication around COVID-19. This is tested with SA Food and Beverage Manufacturing companies and 80% upwards responded that this is already done or being considered with 100% of companies having risk teams and improved COVID-19 communications.

In order to understand company specific mitigation action, the team tests international best practices, as extracted from literature. The key aspects tested includes product management, pricing, staff skills, reviewing planned upgrades, remote work, closure, packaging and online sales, refer to Table 7. Analysis of the FB data indicates four key categories that FB companies have already actioned. These include:

- Adjusting marketing strategies
- Upskilling staff
- Emergency response teams
- New health and Safety Protocols
- Changes in packaging is not a priority of FB business and nether is investing in upgrades.

There is a shift to online sales in all company sizes especially in small business.

Table 7. Analysis of mitigation measures.

	Large						Medium						Micro						Small					
	Already Done	Permanent	Considering	Not applicable	Not considering	I don't Know	Already Done	Permanent	Considering	Not applicable	Not considering	I don't Know	Already Done	Permanent	Considering	Not applicable	Not considering	I don't Know	Already Done	Permanent	Considering	Not applicable	Not considering	I don't Know
Introduced alternative products	11%	0%	11%	11%	21%	0%	22%	0%	5%	20%	24%	2%	33%	0%	0%	0%	0%	0%	11%	0%	15%	37%	15%	0%
Offering lowered prices	14%	0%	7%	7%	14%	11%	22%	0%	20%	7%	20%	2%	0%	0%	0%	0%	22%	11%	19%	0%	7%	15%	33%	4%
Adjusted marketing strategies	25%	0%	18%	0%	7%	4%	37%	0%	24%	0%	12%	0%	22%	0%	11%	0%	11%	0%	37%	0%	33%	7%	0%	0%
Upskilling staff	21%	4%	21%	0%	7%	0%	24%	2%	27%	10%	7%	0%	22%	0%	11%	11%	0%	0%	26%	0%	26%	19%	11%	0%
Emergency response teams	46%	0%	4%	4%	0%	0%	37%	0%	17%	5%	12%	0%	0%	0%	22%	11%	0%	0%	44%	0%	19%	7%	4%	7%
New health and safety protocols	57%	0%	0%	0%	0%	0%	63%	5%	5%	0%	0%	0%	44%	0%	0%	0%	0%	0%	67%	0%	15%	0%	0%	0%
Customer experience improvements (call centres)	25%	0%	4%	4%	11%	7%	22%	0%	15%	22%	12%	0%	11%	0%	11%	11%	0%	0%	15%	0%	11%	26%	22%	0%
Cancelled planned upgrades, expansions or improvements	7%	0%	4%	14%	25%	4%	29%	0%	22%	5%	15%	0%	33%	0%	0%	0%	0%	0%	30%	0%	15%	15%	15%	0%
Sourcing from new suppliers	11%	0%	14%	14%	11%	4%	22%	0%	27%	5%	15%	5%	11%	0%	11%	11%	0%	0%	7%	0%	37%	15%	15%	0%
Online sales	21%	0%	21%	7%	4%	0%	22%	0%	15%	20%	15%	0%	33%	0%	0%	0%	0%	0%	33%	0%	22%	11%	11%	0%
Remote work	36%	0%	14%	0%	7%	0%	22%	0%	17%	17%	17%	0%	33%	0%	0%	0%	0%	0%	19%	0%	19%	19%	26%	0%
Investing in upgrades, renovations or business improvements	11%	0%	29%	4%	4%	4%	12%	0%	27%	10%	20%	2%	11%	0%	11%	0%	11%	0%	7%	0%	15%	15%	37%	0%
Significant downscaling	0%	0%	11%	11%	32%	0%	5%	0%	27%	12%	27%	0%	22%	0%	11%	0%	0%	0%	19%	0%	19%	11%	26%	0%
Temporary closure	0%	4%	0%	21%	29%	0%	10%	0%	10%	12%	37%	2%	11%	0%	11%	0%	11%	0%	11%	0%	11%	15%	41%	0%
Permanent closure	0%	0%	0%	25%	25%	4%	0%	0%	5%	12%	41%	5%	11%	0%	0%	0%	11%	0%	4%	0%	11%	15%	37%	7%
Changes in packaging	4%	0%	21%	11%	11%	7%	10%	0%	20%	10%	29%	0%	11%	0%	0%	11%	11%	0%	0%	4%	33%	11%	26%	0%

3.2.5. Technology

Literature indicates that the drive to mitigate risk to humans can be managed via digital or automation. The assessment of automation is tested in the SA context. Companies are requested to respond to digital supply networks and only the larger companies respond positively at 63% are in support of digital supply chain networks. Training in automation secures a low response rate from companies. Using digital for training has a significantly positive response, as detailed in the training section of this paper.

3.2.6. Climate Change

With the onset of COVID-19, research points to similar events due to the state of global climate, specifically associating health to climate change. Global indications are that due to stimulus packages and reduced costs of borrowing, companies find investment opportunities in COVID-19. Companies are tested with opportunity based on COVID-19 to respond to climate change. Seventy-five percent of SA companies don't find COVID-19 to be an opportunity. Companies do marginally find the need to optimize future investments based on climate; this is at 52% of companies.

3.2.7. Resources Planning

The research team explores the potential for current actions and changes becoming permanent via the question below. *Which of the following do you anticipate becoming a permanent feature in your company post COVID-19?*

The results are summarized in Table 8. Companies strongly support social distancing, transparency, and reconfiguration of the workplace.

Table 8. Analysis of potential new workplace practices.

Permanent Feature	Percentage Response
Elastic workforce (shared workforce)	14%
Healthcare benefits and management	19%
Reallocating workforce to meet business demands	20%
Employees working from home	23%
Collaboration	25%
Programs to encourage virtual working	25%
Reconfiguration of facilities to mitigate COVID-19 spread	34%
Increased communication and transparency	44%
Social distancing at work	55%

The research team focuses on training as a key aspect of data gathering in order to strategize on the FoodBev Manufacturing SETA's response to COVID-19. An initial assessment of the respondents is conducted to establish if companies are recipients of grants. Of the total sample, around 70% of companies responded to the Food and Beverages Manufacturing SETA 2020 grant call.

3.2.8. Training

The first set of questions probe companies' stipend support and continuation of training (Q1/Q2/Q4). The data in Table 9 demonstrates that large and medium companies are better resourced to continue with training during adverse business conditions. Large companies demonstrated higher capacity in payment of stipends, as compared to small and medium companies. Training investments in large and medium companies would enable the SETA to achieve training targets.

Table 9. Summary of training registration and stipends numbers.

	Large		Medium		Micro		Small	
	No	Yes	No	Yes	No	Yes	No	Yes
Does your company have learners registered for training?	1%	25%	15%	22%	1%	4%	17%	14%
Has the company continued training learners during the lockdown period?	8%	17%	27%	11%	4%	1%	20%	11%
Does your company continue to pay stipends to learners during the lockdown?	0%	37%	12%	24%	2%	7%	2%	15%

The next set of questions probes the completion time of the training and the types of training offered (Q3/Q6). In the short-term period of one to three months, the large companies are more disposed to complete the training programs. From the 6-month period, the potential of medium companies completing training substantially increases, with most small companies requiring 12 months. Learnerships, both employed and unemployed, represent 58% of all training programs offered, with internships at 28% and apprenticeships at 13%.

Q11: Envisaged reductions in investments in training and development. Fifty-five percent of all companies responded that a reduction in training investment is anticipated due to the financial challenges being experienced. The 55% comprises 22% of medium companies, 20% of small companies, and 10% of large companies. This implies that medium and small companies are more financially constrained than large companies.

Q7 and Q8: The data indicates 83% of all companies are adopting or planning to adopt a mixed methods approach to skills training. The most commonly adopted tool across all company sizes is video conferencing, followed by online learning programs and resources such as Google classroom (refer to Figure 5. The planned priority methods for adoptions are online learning programs and resources, multimedia including podcasts, and YouTube and virtual reality simulators, respectively, refer to Figure 5.

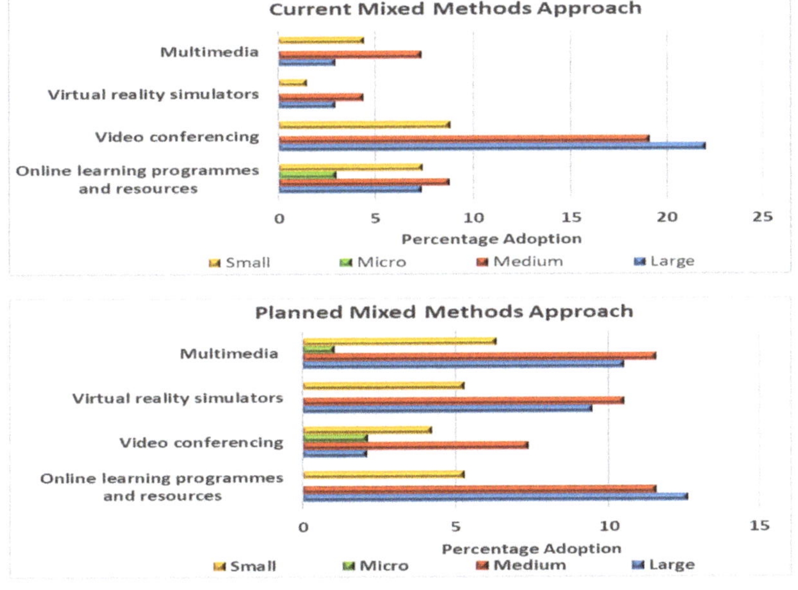

Figure 5. Analysis of current and planned mixed methods for training.

Q9: Challenges experienced by companies and mitigation measures. Ninety-five percent of all companies encountered challenges in continuation of training; with medium sized companies experiencing the most challenges, followed by small and large companies respectively. The critical factors impeding continuity of training across all company sizes in descending order of impact are illustrated in Table 10 and include infrastructure (internet connections, computer resources, etc.), limited digital literacy/skills of users, financial issues, and lack of adapted training programs.

Table 10. Factors impeding continuation of skills training.

	Large	Medium	Micro	Small
Limited digital literacy/skills of users	6.06%	9.85%	0.76%	9.09%
Financial constraints	3.03%	9.09%	1.52%	10.61%
Lack of adapted training programs and resources	5.30%	7.58%	1.52%	4.55%
Infrastructure issues	10.61%	10.61%	2.27%	7.58%

Table 11 illustrates the factors facilitating mixed methods skills development. The companies identified pre-training/online training as the most critical mitigation measure, followed by providing infrastructure as part of our service and investing in program adaption. Providing financial support and providing data were identified as the least important mitigation measures in ensuring continuity of training.

Table 11. Factors facilitating mixed methods skills training.

	Large	Medium	Micro	Small
Pre-training/online training	14%	9%	3%	6%
Providing financial support	1%	6%	0%	4%
Investing in program adaption	6%	6%	1%	3%
Providing infrastructure as part of our services	11%	10%	1%	4%
Providing data	8%	4%	0%	3%

Q12: The final question to companies was on the envisaged areas requiring a change in training. The data identified safety, health, and HR as priority areas across all company sizes. This is in line with current regulations which requires companies to align to social distancing and safety practices at the workplace. Analysis, see Table 12, at the company size level indicates the following training areas as priority:

- Large companies: Categorizes all training areas as important.
- Medium companies: Health, HR, and safety, respectively.
- Small companies: Safety, health, and planning, respectively

Table 12. Business areas requiring training changes.

	Large	Medium	Micro	Small
HR	15%	17%	6%	14%
Supply Chain	14%	13%	17%	10%
Safety	15%	16%	17%	20%
Health	15%	18%	17%	18%
Planning	14%	14%	11%	15%
Automation/Digitalisation	14%	12%	17%	9%
Research and Development	15%	10%	17%	14%

The COVID-19 pandemic has significantly changed the face of business and training. It is apparent that major changes in training, with a shift to digital, is planned and being executed.

4. Discussion

The results provide significant insights into SA Food and Beverage Manufacturing sector alignment with current global trends on COVID-19. The initial reliability tests indicate strong reliability of data. The data indicates a very strong alignment to most categories of COVID-19 responses as extracted from global literature. Essentially, SA FoodBev companies have responded in alignment to global literature trends in most areas, with some exceptions in HR and climate change.

Firstly, and most importantly, on the issue of COVID-19's impact, in the SA context, the analysis indicates a significant number of companies (12 of 106) would likely close with 65% of all companies predicting some future impact of COVID-19, aligning strongly with literature [8,36]. Companies need to structure finances to survive, access government grants, and other government incentives.

COVID-19 has a significant impact on operation, including supply chain. Although around 80% of the companies source and distribute in SA, 50% of those that source out of SA are seeking local sourcing due to supply chain interruptions. Centrally coordinated, digital, supply chain networks are a new global priority [21,23,26,30] and SA companies agree about making digital supply chain skills a priority. SA companies prioritize adjustments in marketing strategies/digital marketing, and upskilling staff and emergency response teams as the top three resource-based mitigations aligning to significant evidence in literature [15,20,27,33,34]. SA companies are not focusing on an elastic workforce to mitigate resources constraints; this is in contrast to global considerations [33,34]. SA companies, most especially medium enterprises, seek technology as a solution with the link between technology and climate change not a fundamental. Specific to operations, the key resources and mitigation actions include social distancing, communication, and facilities reconfiguration. This is in support of the health and safety training required.

Skills infrastructure [16,30] extends beyond the training cycle but extends into the operational space. The cross thematic analysis of all results indicates small and medium enterprises are most significantly impacted (70% of all businesses closing down are small). Small businesses also forecast the highest future impact of COVID-19; this is as per literature [15,36] and anticipated government response. The particular impact on the meat industry is extracted in this study as per literature [6]. An additional and distinct new challenge reflected in the data is that small businesses experience significant additional administration bottlenecks; in the SA context, this seems to be due to the access of government support.

The cross functional analysis is based on various international best practices, as extracted. The first analysis relates to the skills required and delivery mechanisms, a Cronbach Alpha of 0.76 is calculated indicating strong reliability of all skills related questions. As per literature, 4IR skills is a priority and the adoption of 4IR tools to train is an expedited practice under COVID-19. A detailed analysis of the results indicates, in particular, skills required are digital training skills, online curriculum development, and assessment skills [37,38]. The support for setup of Virtual Reality (VR) simulators would provide for the plans to migrate to mixed methods. Infrastructure support is identified through literature [16] and confirmed in this study. Latchem [38] indicates a significant shift to mixed methods of skills development and the data extracted via this study strongly indicates the same with 83% of companies having already adopted or planning mixed methods skills delivery. Asynchronous learning is referenced in literature [37], and the cross functional results is in full agreement. The skills to optimize businesses to survive are essential. Foodbev company's sustainability is skills-based and the Foodbev SETA is most interested in the skills to focus on developing in order to achieve post COVID-19 sustainability.

An analysis of mitigation action provides for a Cronbach Alpha of 0.714, indicating reliability between the various questions asked. SA companies are responding in alignment with international companies with regards to resources-based mitigation, adjusting marketing strategies, upskilling staff, emergency response teams, and new health and safety protocols. Operations-based mitigation include reconfiguration of facilities to mitigate the COVID-19 spread, increased communication, and transparency and social distancing at work as key actions.

5. Conclusions

This research paper identifies global literature relevant to COVID-19 specific to the Food and Beverage Manufacturing sector. A mixed methods approach is adopted to identify global literature and develop a literature relevant instrument to determine the SA contextualized response. The data and statistical analyses provides for insights on the alignment of SA's Food and Beverage Manufacturing Sector with global trends specific to COVID-19. The literature provides insights on current global responses in the Foodbev manufacturing sector. The themes are extracted and developed into a SA specific data instrument seeking SA Foodbev manufacturing companies' responses to COVID-19. A statistical analysis of the instrument data reveals Cronbach Alpha's mostly above 0.6, indicating reliability of the results. The results indicate a strong correlation of the SA Foodbev manufacturing sector in all key responses and mitigations as per global best practice.

Analysis of the results indicates around 10% of companies specified potential closure due to COVID-19, with 65% indicating future impacts due to COVID-19. This is a concerning initial indicator of COVID-19 impact on SME's in SA.

Key mitigation actions in SA include social distancing, communication, facilities reconfiguration and virtual working, this aligns strongly to international best practice. Mitigations not experienced in SA but actioned globally include human resources, specifically exploiting elastic workforces. With the onset of 4IR, technology adjustments are most significant in medium enterprises as a mitigation.

Global literature indicates significant supply chain impacts of COVID-19. SA supply chains are analyzed with 80% of SA companies dependent on the globe for inbound logistics, with 40% for outbound logistics. Forty-seven percent of companies found it difficult to access goods internationally. The most significant response to supply chain challenges being SA companies considering locally coordinated networks and digitalization of supply chains. This aligns strongly to international best practice.

SA Foodbev manufacturing companies have various mitigation measures in place, these align to international literature extractions with the following SA priority ranked responses: (1) adjusting marketing strategies, (2) upskilling staff, (3) emergency response teams, and (4) new health and safety protocols.

The SA specific government response, beyond lockdown support, is skills development to mitigate COVID-19 impact. The SETA is mandated to accelerate COVID-19 mitigation skills development. Literature indicates various global responses to skills development under COVID-19. The data analysis indicates a very strong and rapid response in alignment with global trends. Eighty-three percent of SA companies have already initiated some form of mixed methods skills response.

Priority skills are identified as well as infrastructure support in order to continue to deliver skills training to learners in the sector. Based on this study, the SETA has developed a strategy to support skills development in mitigation of COVID-19. These include digital skills, operational and training skills, supply chain skills, digital marketing skills, and Health Safety Environment (HSE) skills.

Author Contributions: This paper is researched and compiled by three authors with the following responsibilities. Conceptualization: A.T.; methodology: P.M.; A.T.; formal analysis: M.M.; A.T.; investigation: P.M., A.T.; data curation: M.M.; A.T.; writing: P.M.; M.M.; A.T.; original draft preparation: M.M.; A.T.; supervision: A.T. All authors have read and agreed to the published version of the manuscript.

Funding: This research was funded by the FoodBev manufacturing SETA in South Africa.

Acknowledgments: The research team seek to acknowledge the support by the University of Johannesburg for this research project.

Conflicts of Interest: The authors declare no conflict of interest.

References

1. World Health Organization (WHO). Available online: https://www.who.int/ (accessed on 4 April 2020).
2. Johns Hopkins University. COVID-19 Dashboard by the Center for Systems Science and Engineering (CSSE) at Johns Hopkins University (JHU). Available online: https://coronavirus.jhu.edu/map.html (accessed on 3 September 2020).
3. National Institute for Comicable Diseases (NICD). Available online: https://www.nicd.ac.za/covid-19-environmental-health-guidelines/ (accessed on 10 June 2002).
4. Rasool, H. Labour Market Series Part 2 COVID-19, Economy and Labour Market: Reforms for Post-School Education and Training. Available online: https://www.researchgate.net/publication/341423557_LABOUR_MARKET_SERIES_PAPER_TWO_COVID-19_ECONOMY_AND_LABOUR_MARKET_REFORMS_FOR_POST-SCHOOL_EDUCATION_AND_TRAINING (accessed on 10 June 2020).
5. Andersen, K.G.; Rambaut, A.; Lipkin, W.A.; Holmes, E.C.; Garry, R.F. The Proximal Origin of SARS-CoV-2. *Nat. Med.* **2020**, *26*, 45–452. [CrossRef] [PubMed]
6. Norje, C. Economic Impact of Covid-19 on the Meat Industry. Available online: https://sappo.org/economic-impact-of-covid-19-on-the-meat-industry/ (accessed on 30 May 2020).
7. Hyslop, G. Jumping on the Baking Boom: The Bread and Cookie Recipes Hitting the Headlines. Available online: https://www.bakeryandsnacks.com/Article/2020/05/14/Jumping-on-the-coronavirus-baking-boom-The-bread-and-cookie-recipes-hitting-the-headlines?utm_source=copyright&utm_medium=OnSite&utm_campaign=copyright (accessed on 4 June 2020).
8. Food And Agriculture Organization (FAO). *COVID-19 and Smallholder Producers' Access to Markets 2020*; Food and Agricultural Organization (FAO): Rome, Italy, 2020.
9. Citizan, T. Tiger Brands Says Sale of Listeriosis-Tainted Units Will not Affect Class Action. Available online: https://citizen.co.za/business/business-news/2343525/tiger-brands-says-sale-of-listeriosis-tainted-units-will-not-affect-class-action/ (accessed on 5 September 2020).
10. Larkin, P. Government Agree on Transportation of Beer. Available online: https://www.iol.co.za/business-report/economy/sab-government-agree-on-transportation-of-beer-47820667 (accessed on 23 May 2020).
11. Reuters South African Shoppers Stock up on Booze as Sales Resume. Available online: https://headtopics.com/us/south-african-shoppers-stock-up-on-booze-as-sales-resume-13401723 (accessed on 20 June 2020).
12. Hobbs, J.E. Food supply chains during the COVID-19 pandemic. *Can. J. Agric. Econ.* **2020**, *68*, 171–176. [CrossRef]
13. Food and Agricultural Organization (FAO). *Summary of the Impacts of the COVID-19 Pandemic on the Fisheries and Aquaculture Sector, Addendum to the State of World Fisheries and Aquaculture 2020*; Food and Agricultural Organization (FAO): Rome, Italy, 2020.
14. Food And Agriculture Organization (FAO). *How Is COVID-19 Affecting the Fisheries and Aquaculture Food Systems*; Food and Agricultural Organization (FAO): Rome, Italy, 2020.
15. Food And Agriculture Organization (FAO). *COVID-19 and the Risk to Food Supply Chains: How to Respond?* Food and Agricultural Organization (FAO): Rome, Italy, 2020.
16. Food And Agriculture Organization (FAO). *FAO Rolls out Toolkit for Smart Policy Making during the COVID-19 Crisis 2020*; Food and Agricultural Organization (FAO): Rome, Italy, 2020.
17. Falconer, T. Shifting Focus: How COVID-19 is Encouraging Diversification. Available online: https://www.ibisworld.com/industry-insider/coronavirus-insights/shifting-focus-how-covid-19-is-encouraging-diversification/ (accessed on 15 June 2020).
18. Helm, D. The Environmental Impacts of the Coronavirus. *Environ. Resour. Econ.* **2020**, *76*, 21–38. [CrossRef]
19. Rugani, B.; Caro, D. Impact of COVID-19 outbreak measures of lockdown on the Italian carbon footprint. *Sci. Total Environ.* **2020**. [CrossRef]
20. Rosenbloom, D.; Markard, J. A COVID-19 recovery for climate. *Science* **2020**, *368*, 447. [CrossRef] [PubMed]
21. Choudhury, N.R. Food Sector Faces Multipronged Consequences of COVID-19 Outbreak. Available online: https://www.globaltrademag.com/food-sector-faces-multipronged-consequences-of-covid-19-outbreak/ (accessed on 20 May 2020).
22. Aldaco, R.; Hoehn, D.; Laso, J.; Margallo, M.; Ruiz-Salmón, J.; Cristobal, J.; Kahhat, R.; Villanueva-Rey, P.; Bala, A.; Batlle-Bayer, L.; et al. Food waste management during the COVID-19 outbreak: A holistic climate, economic and nutritional approach. *Sci. Total Environ.* **2020**, *742*, 140524. [CrossRef]

23. Weersink, A.; von Massow, M.; Mcdougall, B. Economic thoughts on the potential implications of COVID-19 on the Canadian dairy and poultry sectors. *Can. J. Agric. Econ.* **2020**. [CrossRef]
24. Fitch Solutions. Economic and Suuply Chain Diversification in Asia in the Post Covid-19 Era. Available online: https://www.fitchsolutions.com/country-risk-sovereigns/economics/economic-and-supply-chain-diversification-asia-post-covid-19-era-26-05-2020 (accessed on 15 June 2020).
25. Hossain, S.T. Impacts of COVID-19 on the Agri-food Sector: Food Security Policies of Asian Productivity Organization Members. *J. Agric. Sci. Sri Lanka* **2020**, *15*, 116–132. [CrossRef]
26. Gemmill-Herren, B. Closing the circle: An agro ecological response to covid 19. *Agric. Hum. Values* **2020**, 1–2. [CrossRef]
27. Preiss, P.V. Challenges facing the COVID-19 pandemic in Brazil: Lessons from short food supply systems. *Agric. Hum. Values* **2020**, *37*, 571–572. [CrossRef] [PubMed]
28. Fan, S. Preventing Global Food Security Crisis under COVID-19 Emergency. Available online: https://www.ifpri.org/blog/preventing-global-food-security-crisis-under-covid-19-emergency (accessed on 20 May 2020).
29. World Bank Group. *Assessing the Economic Impact of COVID-19 and Policy Response in Sub Saharan Africa 2020*; World Bank Group: Washington, DC, USA, 2020.
30. Galanakis, C.M. The Food Systems in the Era of the Coronavirus (COVID-19) Pandemic Crisis. *Foods* **2020**, *9*, 523. [CrossRef] [PubMed]
31. World Economic Forum. *The Future of Jobs Report 2018*; World Economic Forum: Cologny, Switzerland, 2018.
32. Kalpana, S.; Priyadarshini, S.R.; Maria Leena, M.; Moses, J.A.; Anandharamakrishnan, C. Intelligent packaging: Trends and applications in food systems. *Trends Food Sci. Technol.* **2019**, *93*, 145–157. [CrossRef]
33. Accenture. Coronavirus Business Economic Impact. Available online: https://www.accenture.com/za-en/about/company/coronavirus-supply-chain-impact (accessed on 15 May 2020).
34. Accenture. Productivity in Uncertain Times Through the Elastic Digital Workplace. Available online: https://www.accenture.com/mu-en/about/company/coronavirus-solution-elastic-digital-workplace (accessed on 15 May 2020).
35. Rizou, M.; Galanakis, I.M.; Aldawoud, T.M.S.; Galanakis, C.M. Safety of foods, food supply chain and environment within the COVID-19 pandemic. *Trends Food Sci. Technol.* **2020**, *102*, 293–299. [CrossRef] [PubMed]
36. Dyal, J.W.; Grant, M.P.; Broadwater, K.; Bjork, A.; Waltenburg, M.A.; Gibbins, J.D.; Hale, C.; Silver, M.; Fischer, M.; Steinberg, J.; et al. COVID-19 Among Workers in Meat and Poultry Processing Facilities—19 States, April 2020. *Mmwr. Morb. Mortal. Wkly. Rep.* **2020**, *69*, 557–561. [CrossRef] [PubMed]
37. Daniel, S.J. Education and the COVID-19 pandemic. *Prospects* **2020**, 1–6. [CrossRef] [PubMed]
38. Latchem, C. *Using ICTs and Blended Learning in Transforming TVET*; United Nations Educational, Scientific and Cultural Organization: Paris, France, 2017.
39. Verawardina, U.; Asnur, L.; Lubis, A.L.; Hendriyani, Y.; Ramadhani, D.; Dewi, I.P.; Darni, R.; Betri, T.J.; Susanti, W.; Sriwahyuni, T. Reviewing online learning facing the Covid-19 outbreak. *Talent Dev. Excell. Spec. Issue* **2020**, *12*, 385–392.

Article

Discussing the Use of Complexity Theory in Engineering Management: Implications for Sustainability

Gianpaolo Abatecola [1,*] and Alberto Surace [2,*]

[1] Department of Management and Law, School of Economics, Tor Vergata University of Rome, 00133 Rome, Italy

[2] CASD—Center for High Defense Studies, Italian Ministry of Defense, 00165 Rome, Italy

* Correspondence: abatecola@economia.uniroma2.it (G.A.); alberto.surace@aeronautica.difesa.it (A.S.)

Received: 19 November 2020; Accepted: 15 December 2020; Published: 19 December 2020 

Abstract: What is the state-of-the-art literature regarding the adoption of the complexity theory (CT) in engineering management (EM)? What implications can be derived for future research and practices concerning sustainability issues? In this conceptual article, we critically discuss the current status of complexity research in EM. In this regard, we use *IEEE Transactions on Engineering Management*, because it is currently considered the leading journal in EM, and is as a reliable, heuristic proxy. From this journal, we analyze 38 representative publications on the topic published since 2000, and extrapolated through a rigorous keyword-based article search. In particular, we show that: (1) the adoption of CT has been associated with a wide range of key themes in EM, such as new product development, supply chain, and project management. (2) The adoption of CT has been witnessed in an increasing amount of publications, with a focus on conceptual modeling based on fuzzy logics, stochastic, or agent-based modeling prevailing. (3) Many key features of CT seem to be quite clearly observable in our dataset, with modeling and optimizing decision making, under uncertainty, as the dominant theme. However, only a limited number of studies appear to formally adhere to CT, to explain the different EM issues investigated. Thus, we derive various implications for EM research (concerning the research in and practice on sustainability issues).

Keywords: complexity theory; engineering management; management; sustainability; conceptual

---

1. Introduction

What is the state-of-the-art literature regarding the adoption of the complexity theory (CT) in engineering management (EM)? What implications can be derived for future research and practices concerning sustainability issues? In EM, addressing these questions through a critical discussion of extant findings is relevant if we consider two, intertwined aspects.

First, in general, the adoption of approaches based on CT has become, in the 21st century, increasingly popular and highly supported. Concerning sustainability related issues, in particular, this is seemingly evident, especially when research grants, funding opportunities, and/or public tenders are released on themes regarding, for example, technology management, open innovation, circular economy, green procurement, or, more generally, sustainable ecosystems [1].

Second, as also highlighted by our analysis in this article, in the 21st century, the use of complexity approaches recurs in decision-making problems, regarding how to improve the effectiveness and efficiency of new product development (NPD), project management (PM), and supply chain management (SCM), or team organization. We know that these aforementioned problems have always been considered as key themes in EM. At the same time, we are confident that, to date, they also represent key challenges towards more sustainable business models [2].

As an example, in addressing a central issue for technology management research, i.e., understanding the nature of the industry environments in which firms play, Ndofor et al. [3] argue that "if the microfoundations of industry environments are indeed strongly impacted by nonlinear relationships, then the industry environment would evolve with chaotic dynamics, as opposed to equilibrium systems" (p. 200). Relatedly, as maintained by McCarthy et al. ([4], p. 437), "early research on NPD has produced descriptive frameworks and models that view the process as a linear system with sequential and discrete stages. More recently, recursive and chaotic frameworks of NPD have been developed, both of which acknowledge that NPD progresses through a series of stages, but with overlaps, feedback loops, and resulting behaviors that resist reductionism and linear analysis."

In the same vein, as stated by Amaral and Uzzi ([5], p. 1034), "a design engineer may know about the reliability of individual parts but find it difficult to estimate how failures in one part of system are tied together or how errors might cascade through the system when apparently separate components have a low probability of failure." Likewise, as posited by Baumann and Siggelkow ([6], p. 116), "should a product design team always consider all components simultaneously, searching for designs that have high overall performance? Or should it first experiment with a subset of components and expand this set gradually in the course of the design process?"

On this premise, starting in the 1960s, several contributions to CT have arisen from various science disciplines, such as biology, mathematics, physics, chemistry, and information technology [7,8]. This is why CT is growing as a cross-disciplinary scientific perspective, offering new approaches and answers, where reductionism demonstrates limits [9,10]. In particular, according to complexity science, the assumption of Newtonian thinking, where everything can be broken down into single pieces, studied separately, and then reassembled to form the initial totality, appears too simplistic when applied to understanding situations characterized by uncertainty and unpredictability [11].

Due to the body of knowledge and continuous, massive expansion of CT, most complexity theorists currently agree on some core characteristics of complexity, and a number of intertwined definitions have been developed over time [12]. Maguire and McKelvey, for example, seminally identify a complex system as "a system (whole) comprised of numerous interacting entities (parts), each of which is behaving in its local context according to some rule(s), law(s) or force(s). In responding to their own particular local contexts, these individual parts can, despite acting in parallel without explicit inter-part coordination nor communication, cause the system as a whole to display emergent patterns—orderly phenomena and properties—at the global or collective level" ([13], p. 4). Likewise, Mitchell conjectures a complex system as a "system in which large networks of components with no central control and simple rules of operation give rise to a complex collective behavior, sophisticated information processing, and adaptation via learning or evolution" ([14], p. 13). Moreover, since complex systems show a tendency to adapt, they are often referred to as complex adaptive systems (CAS); hence, we will use the latter term in this article.

Considering the foregoing, it seems that a conceptual article that critically discusses the current status of complexity research in EM is missing. Thus, the main contribution of our research is that we conceive it as a theoretical start intended to fill this gap. To do so, in Section 2, we first provide readers with the core concepts regarding CT. In Section 3, which constitutes the core of our research, we chose the 21st century to investigate the diffusion of complexity-based accounts in EM. In this regard, we use *IEEE Transactions on Engineering Management (TEM)*, because it is considered as the leading journal in EM [15], and as a reliable, heuristic proxy to start our focus. From this journal, we analyzed 38 representative publications on the topic published since 2000, and went through a rigorous keyword-based article search. Specifically, we provide the pillars of our contribution in terms of key thematic areas investigated and authorship coverage, together with the main research methodologies and core complexity features adopted. Therefore, in Section 4, we discuss some potential (and hopefully valuable) implications of our analysis for sustainability research and practices in this EM field. Section 5 concludes our contribution and presents its limitations.

As a piece of core evidence, our analysis shows that many key features of CAS seem to be clearly observable in the dataset, with modeling and optimizing DM under uncertainty as the dominant theme. Perhaps surprisingly, however, only a limited number of studies still seem to formally adhere to CT, to explain the different EM issues under investigation. This is also why, among the various avenues presented, we suggest that more all-inclusive complexity-based research frameworks would be needed. Accordingly, formally embedding fine-tuned co-evolutionary logics in these frameworks could also add value.

2. Theoretical Background

As previously mentioned, CT represents a multi-disciplinary, modern approach that studies CAS, following its own specific set of laws, behaviors, and characteristics, such as self-organization, and emergence. In principle, CAS can be considered as open systems consisting of several agents locally interacting in a non-linear manner and forming a unique, organized, and dynamic entity; this entity is capable of adapting to, and evolving within, the environment [16]. In other words, CAS have many features in common with living systems; they adapt and evolve through learning.

As mentioned above, a first important characteristic of CAS is the concept of self-organization. The Austrian biologist Von Bertalanffy [17] seminally coins this term in reference to the growth of organisms over time. Self-organizing reflects the ability of CAS to establish an internal organization through adaptation and evolution, without central control.

Relatedly, emergence is a characteristic showed by CAS, where "the behavior of the whole is much more complex than the behavior of its parts" [18] (p. 12). The peculiarity of emergence is that its nature is not necessarily linked to that of the agents [19]. For example, in PM, it has been conjectured that the complex interactions of various parts of a project can generate a specific behavior of the project itself, which can be explained through systemic analysis.

In order to understand how CAS behave, we need to model them, i.e., identify a set of variables that operationally describe these systems. System theory helps with this operationalization [20–22]. In particular, we can define a state variable of CAS as a measurable element of the systems that describes their conditions in a given moment. The state of CAS at a given time is, thus, the set of values held, at that time, by all their state variables [11]. In this regard, there is no formal rule for choosing the appropriate number and type of state variables; however, we can assume that the greater the complexity of CAS (in terms of number of agents and level of interdependence), the greater the variety in type and number of the state variables [13]. Moreover, state variables are represented in an n-dimension space, where n = number of state variables. In this space, each point defines a precise state of the systems (such a state is the state space of CAS). Given a set of state variables, the evolution in time of CAS is a trajectory in its state space [14].

Accordingly, another important characteristic of CAS is that their trajectories in the state space can have three main types of behavior [23]:

1. Order, when the trajectory reaches a point (or an orbit) of the space and then stabilizes. This point or orbit is defined as an attractor. The systems in this regime are stable;
2. Disorder or chaos, when the trajectory shows a chaotic path. In this regime, CAS are completely unstable;
3. Complex regime (or edge of chaos), when the trajectory is attracted by a particular region of the state space. This particular region is known as a strange attractor. In this regime, the systems reach their dynamic equilibrium.

The most interesting type of trajectory appears to be the third (i.e., complex regime), since CAS in this regime show their most relevant behaviors. When CAS reach the complex regime, the conditions are set for all of its peculiarities, i.e., self-organization and emergent behavior, respectively, to be present. However, despite the tendency of the trajectory to orbit around its strange attractor, the evolution of CAS is generally unpredictable [11].

To date, CAS may be found in different contexts, such as economics (e.g., a market), sociology (e.g., a human group), biology (e.g., a cell), business (e.g., an organization), or EM (e.g., a NPD process). In this regard, approaching these contexts through the lens of complexity can, appropriately help face uncertainty and unpredictability [24,25]. In particular, complexity can help model the real world through describing its main characteristics, especially when the deterministic approach seemingly unveils its limits. To do so, to date there are many methodological tools available in the scientific arena. Agent-Based Modeling (ABM), for example, allows simulating the actions and interactions of simple agents, and capturing the emergent and usually complex behavior of the system to which they belong [26]. ABM could also generate adaptive-learning models, which assume that agents have non-linear behaviors, generally based on very simple agent rules [27]. Another tool is fuzzy modeling, which helps face the ambiguity of complexity contexts by introducing un-precise values for the selected variables [28,29]. Likewise, stochastic models countervail the inability to accurately measure well-defined parameters, assuming that an optimal representation may be indeed found within a probability distribution of such measures [30]. Finally, a contribution to help understanding and modeling of complex systems can also be provided by the system of the systems approach [31] because of its tendency to pool resources and capabilities from single systems into a more complex entity, which performs more than the sum of the systems taken separately.

3. Analysis

In order to start discussing the impact of CT on EM, since 2000, after different methodological attempts (in terms of search strings and protocols), we ultimately chose to scan only the *IEEE TEM* journal—considered as the leading journal in the field [15]—through adopting a rigorous keyword-based article search on the EBSCOhost/Business Source Complete research database. In this regard, an initial clarification about the determinants of this methodological choice seems warranted here. This choice happened for two main (intertwined) reasons:

First, at the very beginning of our research project, we attempted to adhere to a traditional systematic review protocol (e.g., [32]). In other words, we initially scanned EBSCOhost/Business Source Complete for all of the articles containing, at least, the keyword "complex*" in their abstract (as known, the asterisk at the end of "complex" allows for different, related suffixes [e.g., complex or complexity]). From a strict procedural view, we are confident that, in principle, this methodological choice would have been, perhaps, more appropriate to initially circumscribing the potentially relevant literature in the field. In practice, however, while performing it, this search produced a large amount of results. These results, in substance, would have made the subsequent steps of a traditional systematic review to be rigorously performed in terms of screening, scanning, evaluating, and selecting, substantively not feasible [33].

We then made various attempts to limit the amount of potentially relevant papers through adding more specific filters, e.g., "engineering management", as keywords in their abstract. However, after making some crash checks through looking at the papers' text, we came to the opinion that this choice would have been too risky, in that it would have probably added opacity to the article inclusion (or exclusion) process. For example, various papers focused on complexity-based innovation, PM, or SCM, thus, in line with the focus of the review, do not contain "engineering management" in their abstract. In other words, at least in our view, this choice would have probably brought the risk of biasing the accountability, rigor, and transparency that is at the core of any systematic review process [34].

Second, as a consequence of the above, we attempted to focus only on *IEEE TEM* to scan EBSCOhost/Business Source Complete for all of the articles containing, at least, the keyword "complex*" in their abstract. This initial step produced 120 results, which then became 111 after eliminating all of the articles published in *IEEE TEM* before 2000 (our focus is on the 21st century), as well as those articles that could not strictly be considered peer-reviewed (e.g., departmental notes or guest editorials). This initial amount of results, we thought, made the subsequent, needed steps for the article inclusion/exclusion, through a rigorous fit for purpose protocol [35] practically feasible.

On this premise, to ensure substantial relevance for our dataset, we scanned all 111 abstracts. Specifically, to be selected: (i) the article abstracts had to formally adopt CT and/or CAS as their theoretical framework; or (ii) if the formal adoption was absent, the presence of the most vivid characteristics of CT had to be clearly identifiable in the abstracts. In particular, as explained in our theoretical framework, this is the case for characteristics such as ABM, emergence, evolutionary dynamics, fuzzy logics, non-linear dynamics, self-organization, stochastic modeling, system of systems, and uncertainty. Overall, this phase reduced our results to 54. Additionally, to ensure conclusive substantial relevance, we repeated this fit for purpose criterion through reading the article texts of all 54 abstracts selected; 38 articles (2000–September 2019) relevant to our research scope finally emerged. In general, this size is consistent with that of many past (e.g., [36]) and recent (e.g., [37]) more traditional systematic reviews, published in the management arena.

In sum, given the exploratory aims of this conceptual article, we believe that, due to the combined mix between the consistency of our dataset and the *IEEE TEM* leading reputation in the EM field [15], an *IEEE TEM*-based initial discussion about the topic coverage can represent: (1) not only a reliable, internationally recognizable, heuristic proxy about the state-of-the-art literature regarding the topic; (2) a (hopefully) challenging starting point to inspire future research efforts in what, as our results show, demonstrates to be a fast-growing, although still not totally conceptually consolidated, area in EM. In this regard, Table 1 synthesizes various, significant items of analysis emerging from our sampled publications. We adapted the thematic areas used in the column "Main Area(s) of Interest" from those present in the ABS 2018 Journal List.

Table 1. An overview of the dataset (in decreasing chronological order per publication year).

N.	Year	(First) Author	Title	Vol., Issue, Pages	Main Area(s) of Interest	Methodology	Industry	Complexity Characteristics	Main Content
1	2019	De	Multiobjective approach for sustainable ship routing and scheduling with draft restrictions	66, 1, 35–51	Operations	Non-dominated Sorting Genetic Algorithm	Maritime Transportation	Evolutionary, Non-linear	Through considering different variables relevant in maritime transportation, the work provides a genetic algorithm to support complex decisional processes in this industry.
2	2019	Li	Optimizing the labor strategy of a professional service firm	66, 3, 443–458	Human Resource Management	Labor Strategy Optimization	Global Professional Services	Stochastic, Non-linear, Uncertainty	Through non-linear analysis and uncertainty modeling, the work designs the Labor Strategy Optimization framework to strategically optimize the use of workforce at firm level.
3	2019	Yu	A complex negotiation model for multi-echelon supply chain networks	66, 2, 266–278	Operations	Simultaneous Multi-attribute Multi-item Modeling	-	ABM	Through using ABM, the work proposes a framework to support complex negotiation problems in supply chain networks.
4	2018	Ndofor	Chaos in industry environments	65, 2, 191–203	Strategy	BDS Test, Correlation Dimension Test, Lyapunov Exponent	Various	CAS, Non-linear	The work uses the nonlinear dynamical system methods from CT to study how different industry environments evolve over time.
5	2017	Liu	Novel two-phase approach for process optimization of customer collaborative design based on fuzzy-QFD and DSM	64, 2, 193–207	Innovation	Design Structure Matrix, Quality Function Deployment	Automotive	Fuzzy, Uncertainty	In the context of NPD, the work uses design structure matrix and quality function deployment to propose a two-stage model focused on customer satisfaction and cooperation.
6	2016	Geng	A new fuzzy process capability estimation method based on Kernel function and FAHP	63, 2, 177–188	Operations	Process Capability Indicators/Kernel Function	Process	Fuzzy, Non-linear, Uncertainty	Through a simulation in the Tennessee Eastman process, the work proposes a new method to estimate the production process capability, together with a new criterion for the evaluation of capabilities and performance.

Table 1. *Cont.*

N.	Year	(First) Author	Title	Vol., Issue, Pages	Main Area(s) of Interest	Methodology	Industry	Complexity Characteristics	Main Content
7	2016	Giannoccaro	Examining the roles of product complexity and manager behavior on product design decisions: An agent-based study using NK simulation	63, 2, 237–247	Innovation	NK Model	-	ABM, CAS, Evolutionary	Through a methodology drawn from complexity science, the work studies what behavioral factors can influence project managers in their choice regarding the degree of centralization of decisions about product design.
8	2016	Sarker	Internal visibility of external supplier risks and the dynamics of risk management silos	63, 4, 451–461	Operations	Bounded Rationality, Contingency Theory	Manufacturing	Non-linear, Uncertainty	The work uses bounded rationality and contingency theory to explain non-linear and non-deterministic perception of risks associated with the SCM process.
9	2016	Zhang	A stochastic ANP-GCE approach for vulnerability assessment in the water supply system with uncertainties	63, 1, 78–90	Operations	Analytical Network Process, Game Cross Evaluation	Water Supply	Fuzzy, Non-linear, Stochastic, Uncertainty	Through the case study of the Shanghai water supply system, the work proposes a stochastic multi-criteria approach for the vulnerability assessment of each component in the system.
10	2015	Herrmann	Predicting the performance of a design team using a Markov chain model	62, 4, 507–516	Innovation, Human Resource Management	Markov Chain	Motors	ABM, Stochastic	Proposing a Markov chain model, the work studies when it is convenient for bounded rational problem solvers/agents in search for an optimal solution, to decompose complex problems of product development in different, less complex sub-problems.
11	2015	Jiang	Optimizing cooperative advertising, profit sharing, and inventory policies in a VMI supply chain: A Nash bargaining model and hybrid algorithm	62, 4, 449–461	Operations	Non Linear Nash Bargaining Model/Hybrid Algorithm	Retail	Evolutionary, Non-linear, Stochastic	The work develops a Nash bargaining and hybrid algorithm model to optimize the complex joint DM regarding vendor managed inventory supply chains.

Table 1. *Cont.*

N.	Year	(First) Author	Title	Vol., Issue, Pages	Main Area(s) of Interest	Methodology	Industry	Complexity Characteristics	Main Content
12	2015	Parraguez	Information flow through stages of complex engineering design projects: A dynamic network analysis approach	62, 4, 604–617	Information Management, Innovation	Dynamic Network Analysis	Renewable (bio-mass) Energy	Emergence, Evolutionary	Through the dynamic network model developed, the work offers a tool to dynamically quantify and analyze the information flows among the activities of complex engineering design projects.
13	2015	Tsilipanos	Modeling complex telecom investments: A system of systems approach	62, 4, 631–642	Strategy	Genetic Algorithm	Telecom	Emergence, Evolutionary, Stochastic, System of Systems, Uncertainty	Focusing on telecommunications, and adopting the system of systems method from CT, the work models a genetic algorithm useful to study optimal DM and budget allocation.
14	2015	Villalba-Diez	Improving manufacturing performance by standardization of interprocess communication	62, 3, 351–360	Operations, Information Management	Interprocess Communication Holon	Engine Manufacturing	CAS	Studying the standardization of interprocess communication in complex supply chain networks, the work proposes a holistic model from which the manufacturing performance can increase.
15	2014	Kaki	Scenario-based modeling of interdependent demand and supply uncertainties	61, 1, 101–113	Operations	Scenario Based Modeling	Manufacturing	Non-linear, Stochastic, Uncertainty	Through the case of a manufacturing company, the work develops a scenario-based framework for modeling the interdependence between demand and supply uncertainties.
16	2014	van de Kaa	Supporting decision making in technology standards battles based on a fuzzy analytic hierarchy process	61, 2, 336–348	Strategy	Analytic Hierarchic Process	Technology Standards	Fuzzy, Emergence, Uncertainty	The study uses a fuzzy analytic hierarchic process to model the emergence, selection, and survival of technological standards over time.

Table 1. *Cont.*

N.	Year	(First) Author	Title	Vol., Issue, Pages	Main Area(s) of Interest	Methodology	Industry	Complexity Characteristics	Main Content
17	2012	Muller	Relationships between leadership and success in different types of project complexities	59, 1, 77–90	Management Science, Human Resource Management	Leadership Dimensions Questionnaire	Project Management Institute/ International Project Management Association	CAS, Uncertainty	The work studies whether project complexity moderates the relationship between the leadership competences of project managers and project success.
18	2012	Shafiei-Monfared	Fuzzy complexity model for enterprise maintenance projects	59, 2, 293–298	Management Science	Graph Complexity Model	Aircraft Engines	Fuzzy, Uncertainty	Through fuzzy modeling, the work defines different levels of project (managerial and/or technical) complexity, with the model useful for budgeting, planning, and resource allocation.
19	2012	Stryker	Creating collaboration opportunity: Designing the physical workplace to promote high-tech team communication	59, 4, 609–620	Information Management, Human Resource Management	Hierarchical Regression Analysis	Pharmaceutical	Uncertainty	The work studies how the probability to achieve complex team tasks is impacted by the relationship between the physical design of the workplace and the face-to-face communication among team members.
20	2012	Van der Vooren	Managing the diffusion of low emission vehicles	59, 4, 728–740	Strategy	Modeling based on vehicle technologies, infrastructures, and consumers	Automotive	ABM, Non-linear, Stochastic	Through using ABM, the work studies the competition for technological standards between a number of low emission vehicle technologies and the dominant fossil fuel based.
21	2011	Mikaelian	Real options in enterprise architecture: A holistic mapping of mechanisms and types for uncertainty management	58, 3, 457–470	Management Science	Real option analysis	Surveillance	Emergence, Evolutionary, Uncertainty,	The work develops a holistic approach based on real option analysis to manage flexibility and DM under uncertainty in complex engineered systems.

Table 1. *Cont.*

N.	Year	(First) Author	Title	Vol., Issue, Pages	Main Area(s) of Interest	Methodology	Industry	Complexity Characteristics	Main Content
22	2011	Revie	Supporting reliability decisions during defense procurement using a Bayes linear methodology	58, 4, 662–673	Operations, Management Science	Bayes Linear Modeling	Defense	Uncertainty	Through an industrial application in a defense procurement project setting, the work proposes a Bayes linear methodology to support the reliability of DM.
23	2011	Tripathy	Organizing global product development for complex engineered systems	58, 3, 510–529	Innovation, International Business	Design Structure Matrix	Various	CAS	Adopting the perspective of complex engineered systems, the work models the offshoring and onshoring of the activities associated with NPD at global level.
24	2010	Goh	Uncertainty in Through-Life costing—Review and perspectives	57, 4, 689–701	Management Science	Through-Life Cost	-	Fuzzy, Uncertainty	The work reviews how uncertainty is classified in the engineering literature and how it is conceived in the through-life cost estimation methodology.
25	2010	Zhang	An optimal-control-based decision-making model and consulting methodology for service enterprises	57, 4, 607–619	Management Science	Approximate Dynamic Programming Algorithm	Service	Fuzzy	In the context of service management, the work proposes a DM model based on an approximate dynamic programming algorithm, which can be useful to manage the planning and evaluation of complex projects.
26	2009	Levardy	An adaptive process model to support product development project management	56, 4, 600–620	Innovation, Management Science	Adaptive Product Development Process Modeling	Packaging	ABM, CAS, Evolutionary Fuzzy, Stochastic, Uncertainty	The study conjectures the process of product development as a CAS, featured by a general class of self-organizing activities/rules, able to adapt to the changing state of the process.
27	2008	Jun	A modeling framework for product development process considering its characteristics	55, 1, 103–119	Innovation	Modeling based on product development characteristics	Automotive, Electronics	CAS, Evolutionary, Uncertainty	The work provides modeling patterns for the product development process based on its iterative, evolutionary, uncertain, and cooperative characteristics.

Table 1. *Cont.*

N.	Year	(First) Author	Title	Vol., Issue, Pages	Main Area(s) of Interest	Methodology	Industry	Complexity Characteristics	Main Content
28	2007	Pathak	On the evolutionary dynamics of supply network topologies	54, 4, 662–672	Operations	Modeling based on supply network topologies	Automotive	ABM, CAS, Emergence, Evolutionary, Stochastic, Uncertainty	Through combining CAS with industrial growth, network, market structure, and game theories, the work investigates how supply network structures evolve and survive over time.
29	2007	Raisinghani	Strategic e-business decision analysis using the analytic network process	54, 4, 673–686	Strategy	Analytic Network Process	E-Business	Non-linear	In the context of e-business, the work uses the analytic network process to model optimal DM when decision complexity increases.
30	2006	Batallas	Information leaders in product development organizational networks: Social network analysis of the Design Structure matrix	53, 4, 570–582	Information Management, Human Resource Management, Innovation	Social Network Analysis/Design Structure Matrix	Aircraft Engines	CAS, Non-linear	In settings featured by the complexity of product development projects, the work uses social network analysis to model and evaluate the information flow, with a focus on the identification of information leaders.
31	2005	Cho	A simulation-based process model for managing complex design projects	52, 3, 316–328	Innovation, Management Science	Design Structure Matrix Simulation Modeling	Aerospace	Stochastic, Uncertainty	Through an industrial application in the aerospace industry, the work uses the design structure matrix modeling to propose an approach for managing complex product design projects.
32	2005	Jun	On identifying and estimating the cycle time of product development process	52, 3, 336–349	Innovation	Modeling based on product development characteristics	Automotive	Evolutionary, Stochastic	The work provides modeling patterns for the product development process based on its characteristics of interaction, evolution, and uncertainty.
33	2005	Williams	Assessing and moving on from the dominant project management discourse in the light of project overruns	52, 4, 497–508	Management Science	Systemic Modeling	-	CAS, Emergence, Stochastic, Uncertainty	Through reviewing PM theories, the work proposes systemic modeling as a useful, learning based approach to manage the uncertainty and emergence characteristics associated with complex projects.

Table 1. *Cont.*

N.	Year	(First) Author	Title	Vol., Issue, Pages	Main Area(s) of Interest	Methodology	Industry	Complexity Characteristics	Main Content
34	2004	Lin	New product go/No-go evaluation at the front end: A fuzzy linguistic approach	51, 2, 197–207	Innovation	Logic-Based Screening Model/Linguistic Multi-criteria Decision	Machinery	Fuzzy, Uncertainty	Through an industrial application on a new machining center, the work proposes a new screening model based on fuzzy logics and linguistic approximation to assess the design of new products.
35	2004	Sia	Effects of environmental uncertainty on organizational intention to adopt distributed work arrangements	51, 3, 253–267	Human Resource Management	Partial Least Square	Various	Uncertainty	The work is an exploratory study about the convenience of using distributed working arrangements as an organizational innovation to face environmental uncertainty.
36	2004	Xirogiannis	Fuzzy cognitive maps in business analysis and performance-driven change	51, 3, 334–351	Management Science	Cognitive Mapping	Financial Sector	Evolutionary, Fuzzy, Non-linear	Positioned in the business process reengineering area, the work uses fuzzy cognitive mapping to analyze performance-driven reengineering processes.
37	2003	Huntley	Organizational learning in open-source software projects: An analysis of debugging data	50, 4, 485–493	Information Management	Adaptive Learning, Debugging	Open Source Software	Non-linear, Uncertainty	The work studies open-source debugging as a form of organizational learning, with a focus on the open source approach as a hedge against system complexity.
38	2002	Vachon	An exploratory investigation of the effects of supply chain complexity on delivery performance	49, 3, 218–230	Operations	Modeling based on supply chain characteristics	Textile, Machinery	Stochastic, Uncertainty	The work provides a conceptual model characterizing the complexity features of a supply chain, which is useful to understand the linkage between SCM and delivery performance.

Source: own elaboration.

In the four sub-sections below, we analyze these items per key content lines.

3.1. Themes

In terms of fields, as a premise, we can consider about two-thirds of our sampled publications as falling into traditional EM, one-third into technology management, and substantially none in emerging technologies (Figure 1).

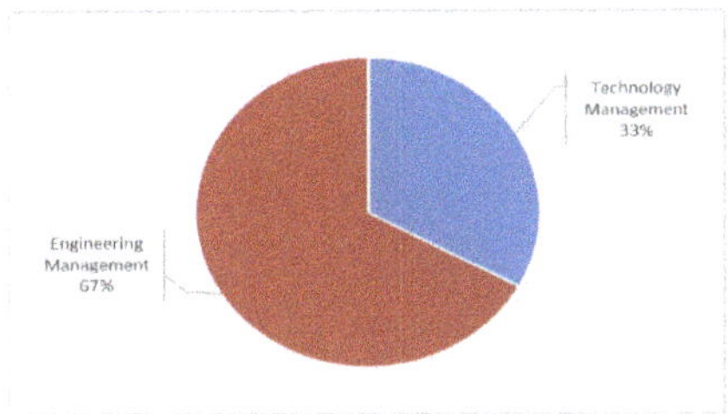

Figure 1. Publication coverage per field. Source: own elaboration.

In more detail, as Figure 2 shows, since 2000 CT has been associated with a wide spectrum of topics and themes associated with the fields above.

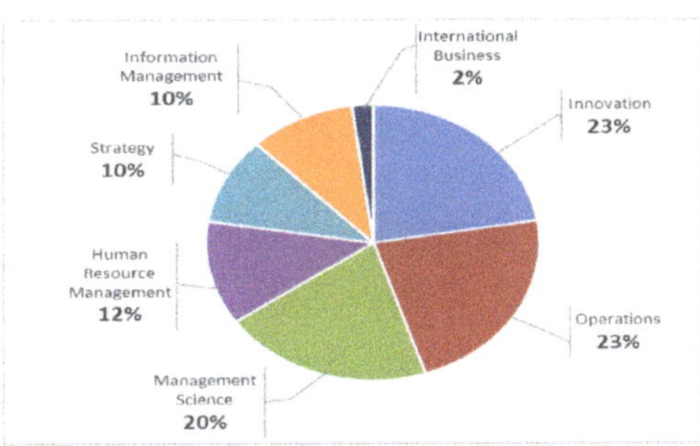

Figure 2. Publication coverage per key thematic areas. Source: own elaboration.

In particular, as Figure 2 shows, innovation, operations, and management science represent, as we could somehow expect, the most investigated areas. In this respect, works on the use of CT in DM processes, regarding NPD, procurement, and supply chain, or PM, specifically prevail. Interestingly, at the same time, considerable (although minor) amounts of observations fall into the areas of human resource management, strategy, and information management. In this instance, for example, the focus is on the use of CT to increase team productivity, competitive capabilities in (technological) environments, or the efficiency/effectiveness of intraorganizational communication.

3.2. Timely Distribution and Authorship

As Figure 3 illustrates, the time distribution of the publications witnesses an increase, especially if we separate the articles published in the years between 2000 and 2010 from those published between 2011 and 2019.

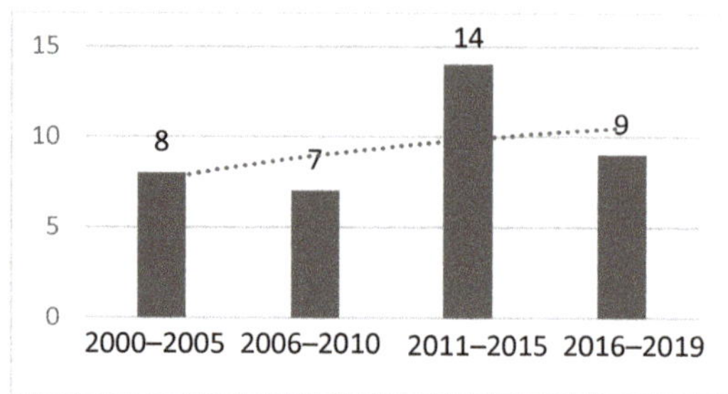

Figure 3. Evolving trend of the publications. Source: own elaboration.

On this premise, interesting evidence seemingly emerges if we focus on various features regarding the authorship coverage of our sampled publications (Figures 4 and 5).

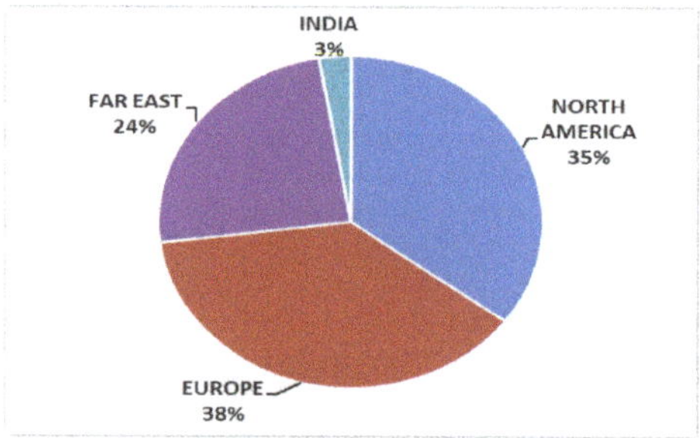

Figure 4. Publication coverage per geographical source. Source: own elaboration.

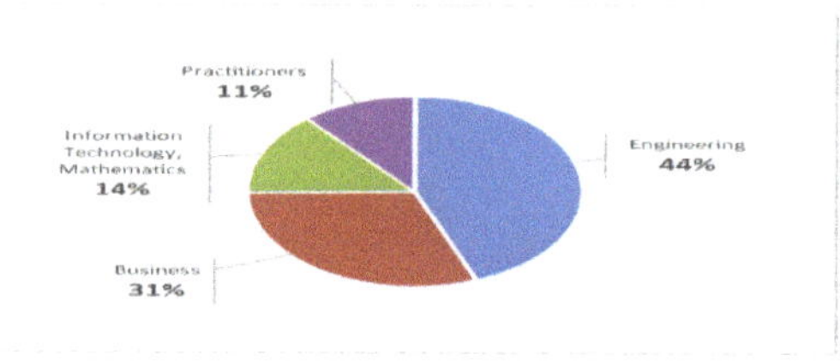

Figure 5. Publication coverage per author affiliation. Source: own elaboration.

Figure 4 substantially shows what we could consider the geographical source of our sampled publications. In particular, we developed this data-driven figure by contemporaneously considering: (1) the first author (N = 37, net of duplicates) of each publication; (2) the country in which s/he was awarded her/his PhD. In this regard, we chose to specifically focus on first authors because of the internationally acknowledged leadership role, which, in general, any first author has in terms of the research design of a publication. At the same time, we preferred to focus on the country in which the first authors were awarded their PhD rather than on their strict nationality because we thought the former could represent a more reliable proxy for the cultural orientation (and associated approach) towards the topic.

Having clarified the above, as shown in Figure 4, the geographical source of our dataset appears substantially balanced between Europe and North America, followed, at the same time, by a significant presence of Far East countries (e.g., China, Japan, Taiwan, Hong Kong, Singapore, and South Korea).

Correspondingly, Figure 5 shows the publications' coverage by author affiliation. In this case, we developed this data-driven figure by considering all of the authors (N = 107, net of duplicates) in our dataset. Interestingly, as shown in the figure, engineering schools/departments prevail, but business schools/departments also occupy a significant portion. at the same time, although in minor percentages, Figure 5 also evidences the presence of scholars from other schools/departments, such as information technology or mathematics, and practitioners as well. We could argue that this evidence can be interpreted as consistent, as explained in our theoretical framework, with the multidisciplinary nature of the approaches to CT.

3.3. *Methodologies, Settings, and Complexity Features*

Almost all of the studies are based on conceptual, mathematical modeling, with the vast majority also tested through industrial applications, relying, for the largest part, on quantitative methods. Interestingly, on the one hand, the conceptual modeling is featured by a wide range of techniques, these varying, for example, from genetic algorithms to design structure matrices, or analytical hierarchical/network processes. At the same time, on the other hand, many of these techniques share the common feature of grounding on fuzzy logics, stochastic modeling, or ABM as their basis. From more than one aspect, similar highlighting can also regard the context of the industrial applications. In fact, the general settings are heterogeneous ranging, for example, from aerospace, to automotive, manufacturing, or services. However, almost all of these settings share a strong hi-tech component in what is specifically observed.

Figure 6 expands on Table 1, offering statistics about the presence of the inner complexity characteristics in our dataset. In particular, we built this figure through the assumption that more than one characteristic can be simultaneously present in the observed publications.

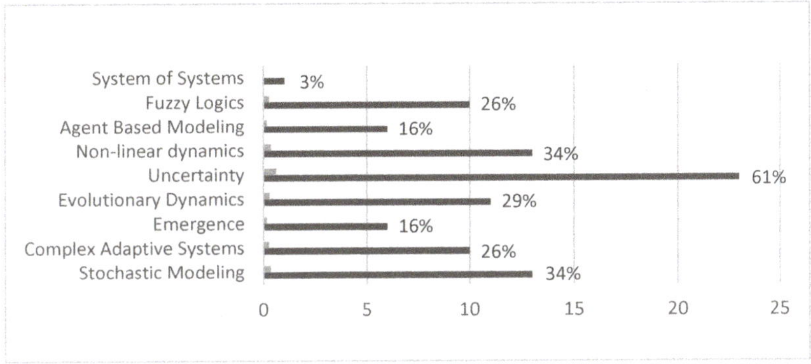

Figure 6. Presence of the complexity characteristics in the dataset. Source: own elaboration.

As evidenced in Figure 6, the study of DM and problem solving under uncertainty (and how to manage it) largely prevails, and generally serves as the ground basis for various lines of inquiry, with one or more complexity characteristic often contemporaneously present with uncertainty itself. In particular, as evidenced in the figure, uncertainty is frequently associated with non-linear dynamics and/or, as previously mentioned, stochastic modeling. The former, for example, is interestingly highlighted by Xirogiannis and Glykas [38] in their study on how performance-driven business reengineering processes work and how they could eventually work better. The latter, in parallel, is used more than once to provide insight on how to model the complexity, towards efficiency and effectiveness, regarding NPD, PM practices, or SCM.

An interesting number of observations also include the use of fuzzy logics in conjunction with uncertainty. In the area of management science, for example, and with a focus on PM, Shafie-Monfared and Jenab [39] use fuzzy modeling to identify different degrees of project complexity, based on the differentiation of managerial and technical features. Their framework can usefully provide support to budgeting, planning, and resource allocation. Similarly, through the case study of a new machining center, Lin and Chen [40] propose a new method to evaluate new product design, based on fuzzy logics in general, and linguistic approximation in particular.

Finally, in our dataset, uncertainty is also repeatedly associated with evolutionary dynamics. For example, Mikaelian et al. [41] develop a holistic, evolutionary approach, based on real option analysis, to manage flexibility and DM under uncertainty. In a similar vein, Giannoccaro and Nair [42] heavily rely on complexity science and evolutionary mechanisms to study what (and how) behavioral traits of project managers can shape their decisions regarding product design.

3.4. Complexity-Based Evidences

In relation to the third issue analyzed above, however, it seems that only a limited number of studies still formally adhere to the lenses of CT, and/or CAS, to explain the different EM issues under investigation. For example, in the innovation area, Tripathy and Eppinger [43] focus on complex engineered systems, with particular regard to the offshoring and onshoring activities associated with NPD at a global level. In detail, they use five case studies from electronics, equipment, and aerospace to study the complexity of the interactions between the product and process structures, and the strategies planned and implemented at firm level. On the basis of their findings, these scholars then propose theoretical trajectories aimed at improving the DM configuration regarding global product development in complex engineered systems. As their core idea, the modularity in design and development should be separated from that in manufacturing; furthermore, the development of the system architecture, which is a core capability, should not be offshored.

In a similar vein, Levardy and Browning [44] conjecture the processes of NPD as CAS. These scholars oppose linear, time-based vertical scheduling, in that they theorize these processes as featured by a general class of activities/rules, which can self-organize and adapt to their changing state. The implications of their modeling for DM in EM are interesting; in fact, their adaptive model considers product development as a DM process, in which each decision is potentially able to maximize the expected value of the overall project based on the particular state, in any given moment, of its internal and external variables.

Again, in the context of NPD, the work by Jun and Suh [45] appears particularly worth of explanation. They also provide a theoretical framework for the process, composed not only of iterative but also evolutionary, uncertain, and cooperative characteristics. Through an industrial application in the automotive, electronics, and environmental settings, their modeling demonstrates its potential utility to engineers and project managers involved in planning, organizing, and monitoring the design and implementation of new product initiatives.

Following the above evidences about innovation, in the strategy area, Ndofor et al. [3] use the nonlinear, dynamical system methods from CT to study how different industry environments evolve over time. In particular, adopting three operationalizations, classically utilized to discover nonlinear variable dynamisms, these scholars evidence that many industries evolve in a chaotic regime, where uncertainty increases proportionally to hypercompetitive settings. Similarly, Tsilipanos et al. [30] analyze investments in the telecommunication industry through using a methodological approach typical of CT. Specifically, these scholars model this industry as a system of systems, and use the MATLAB software to create a genetic algorithm able to provide results based on stochastic, emergent modeling. Tested through an industrial application, the more general value of their modeling, also in terms of implications for EM, mainly consist of the possibility to provide prospective investors with theoretical support to efficient DM and budget allocation.

Finally, in the operations and supply chain area, the research by Pathak et al. [46] seemingly deserves attention. Through combining the CAS approach with industrial growth, networks, market structure, and game theories, these scholars investigate how supply network structures can evolve and survive over time. The observations from their agent-based study in the U.S. automotive industry can be of particular appeal to engineers. Specifically, they find that the type of environment and the speed of adaptability both affect the survival chances of supply networks; in peaceful settings, on the one hand, the topological evolution of the networks is relatively stable, with centralized or linear network structures, often able to guarantee survival over the long term. In more competitive settings, however, only the hierarchical structure seems able to provide networks with adequate long-term survival chances.

4. Discussion and Implications

In this conceptual article, we have focused on the adoption of CT in EM since 2000. At the beginning, we introduced the key conceptual pillars of CT (and CAS). Subsequently, because of its status of being a leading journal in the field, we chose *IEEE-TEM* as a reliable, heuristic proxy to analyze and discuss those publications formally, and/or substantially, referring to complexity approaches. Therefore, we can synthesize the results from our analysis into the following three main evidences.

First, from 2000, the adoption of CT in EM has been associated with a wide range of key themes in the field. NPD, SCM, and PM prevail. At the same time, a considerable number of observations also regards team productivity, competitive capabilities in (technological) environments, and intraorganizational communication.

Second, this adoption was seen in an increasing amount of publications, especially if we consider the years 2011 to 2019. Conceptual modeling developed through a wide-range of techniques largely prevails in our dataset, then quantitatively tested in various (almost hi-tech based) industrial settings. This, again, also appears in line with the plurality and heterogeneity of analytical tools and (high-tech) settings traditionally employed in EM [47]. At the same time, the common feature among these techniques is that they are mostly based on fuzzy logics, stochastic modeling, or ABM.

Third, many key ingredients of CT seem to be quite clearly observable in the analyzed publications. Accordingly, modeling and optimizing DM under uncertainty results as the dominant theme; this theme, at the same time, is not only often associated with the mentioned fuzzy logics, stochastic modeling, and ABM, but also with non-linear and/or evolutionary dynamics. Perhaps surprisingly, however, only a limited number of studies still seems to formally adhere to CT to explain the various EM issues under investigation.

From what is summarized above, some implications for EM (concerning the research in and practice on sustainability issues) can also be derived. These implications are exposed below, sequentially ordered per item of focus.

4.1. Areas of Investigation and Leadership

Regarding the areas of investigation, on the one hand, as previously written, our results show good coverage of complexity-based approaches in key EM areas. On the other hand, we think that additional areas could also become objects of research in this field. For example, the emerging technologies/technology intelligence area could be expanded through complexity-based observations concerning artificial intelligence or Internet of Things. In fact, on both of these topics, we could not find any evidence in our analysis. Moreover, further studies could also look into how to develop, from engineers to leaders; correspondingly, we could find good coverage of human resource management in general, but, apart from scant exceptions, we could not find sizeable evidence about complexity-based leadership [48] in our analysis.

Regarding the above, for instance, and with a focus on the potential impact of complexity-based leadership on the effectiveness and efficiency of innovation (e.g., NPD) and change, the recent work by Burnes [49] appears remarkable. In particular, according to this scholar (p. 84), "unless employees have the freedom to act as they see fit, self-organization will be blocked, and organizations will die because they will not be able to achieve continuous and beneficial innovation." Furthermore, he states (p. 84), "neither small-scale incremental change nor radical transformational change works: instead, innovative activity can only be successfully generated through the third kind of change, such as new product and process development brought about by self-organizing teams."

Relatedly (p. 84), "because organizations are complex systems, which are radically unpredictable and where even small changes can have massive and unanticipated effects, top-down change cannot deliver the continuous innovation which organizations need in order to survive and prosper. Instead, it is argued that organizations can only achieve continuous innovation if they position themselves at the edge of chaos". According to Burnes, self-organization is the only way to reach and keep this position, and is itself based on rules that are order-generating. The key point here is that, if the latter (i.e., rules) result in no longer fitting the organizational context, they can be re-created exactly because of the existence of the former (i.e., self-organization).

Having explained the above, a noteworthy example of complexity-based leadership can be offered by a recent case study considering a military organization as a CAS [50], with a focus on its inner complex dynamics, as an enabler to increase organizational effectiveness. As the case demonstrates, despite the traditionally hierarchical and linear characteristics of military organizations, in order to face the surrounding complexity, the rapidly changing defense environment has substantially proved to need a more adaptable and flexible structure.

On this basis, the military leader willing to adopt a complex approach to the commanding action will seek to foster those dynamics typical of CT (such as non-linear relationships and feedback) in order to increase adaptability and organizational learning. This also implies the need to drive the organization from hierarchical to network-centered dynamics, thus assuring governance cohesion throughout the organization, thanks to the development of a shared vision across the top management team. In principle, this perspective can also be considered as presenting similarities with many conceptual underpinnings featuring the notion of socio-technical systems (e.g., [51]).

4.2. Settings of Observation and Research Methodologies

Concerning the settings of observation, in a similar vein as above, we could argue that, together with the key high-tech contexts in EM already emerging from our analysis, other central contexts in the sustainability field, such as energy, healthcare, and construction, could become the basis of complexity-based observations. Regarding these contexts, in fact, apart from a few exceptions our analysis could not evidence any specific focus.

Relatedly, with respect to research methodologies, on the one hand, our findings have shown that conceptual modeling tested through quantitative techniques has largely prevailed in the complexity-based observations in EM. On the other hand, however, we maintain that designing and conducting in-depth qualitative case studies [52] should also be important in the field. In this regard, (a) we are substantially in line with those scholars [53,54] who have, for a long time, generally argued that case studies are highly appropriate in complementing computational methods to understand the distinctive features of CAS; and (b) we are particularly in line with those scholars who have used the properties of case studies to develop complexity-based observations in key EM fields, such as NPD.

Taking the above into account, for example, McCarthy et al. [4] used a comparative analysis of three cases to examine how the CAS features of non-linearity, self-organization, and emergence can occur in NPD processes. In particular, these scholars conceive a model of NPD processes, as CAS, featured by three levels of DM, in stage, review, and strategic, respectively. Taking a middle ground between stage gate, chain linked, and chaotic models of NPD, their analysis produces interesting results. In their view, NPD is not necessarily a fixed process; it can adapt and switch from linear to chaotic (and vice versa), thus producing corresponding degrees of incremental or radical innovation. In the practice of EM, their model would be very helpful to avoid the DM traps, potentially regarding the search for fit between (new) product, (new) process, and market demand.

4.3. Conceptual Frameworks

Our analysis has shown that, among the many key ingredients of CT quite clearly observable in the analyzed publications, modeling and optimizing DM under uncertainty appears to prevail. Accordingly, we support the recent argument by Baumann and Siggelkow [6] that, in conditions of rationally bounded problem solving, understanding whether integrated (i.e., entirely and simultaneously performed) or chunky (i.e., incrementally expanded) search processes are the most appropriate could also add value. Again, in a technology innovation context of NPD, these scholars focused on this issue through the application of a simulation model. Their analysis has evidenced interesting results: incremental should be preferred to integrated patterns of search when time pressure is not a variable under consideration; moreover, the larger the chunks added at the beginning of the search process, the less the need of a totally incremental search.

According to our results for EM, complexity-based observations have often associated the uncertainty variable with fuzzy logics, stochastic modeling, and ABM, but also with non-linear and/or evolutionary dynamics. As this association has mostly happened on a separate basis (see Table 1), we argue that all-inclusive, complexity-based frameworks could be developed further. Again, this claim corresponds with other key evidence from our analysis: as previously stated, we have shown that, in EM, only a limited number of studies still seem to formally adhere to CT to explain the EM issues under investigation.

The more comprehensive frameworks elicited above could then be tested in different EM settings to assess their reliability. For example, a recent, remarkable attempt of this kind has been the Generalized Complexity Index developed by Jacobs [55]. Based on the three dimensions of multiplicity, diversity, and interconnectedness, this index can be used as an analytical decision tool to evaluate the pros and cons of potential portfolio diversification and/or product differentiation. Furthermore, especially in these learning-based, innovation contexts, distinguishing between complex adaptive and complex generative systems [56] could also be valuable. While the former systems are able to adapt without the need for radical changes, the latter can witness changes which largely modify their inner features and even generate new entities.

4.4. Co-Evolutionary Dynamics in Complexity-Based Research Designs

The issue of the interconnectedness brings us to the last item to be discussed in terms of potential implications for sustainability, which is a direct call to embed more fine-tuned co-evolutionary perspectives in complexity-based research designs [57,58]. Specifically, we argue, this call appears to have particular momentum if (and when) hypercompetitive technology environments are under investigation. In fact, recalling what was recently demonstrated by Ndofor et al. [3] on the basis of their 36-year observations of 19 industry sectors, these environments are often chaotic, i.e., featured by a significant degree of a non-linear relationship among elements, together with inter and path dependence. As a fast growing meta-theoretical perspective in social sciences [59–62], and being generally conceived as the joint and dynamic outcome between industry, managerial, and environmental forces [63–65], co-evolution demonstrated effectiveness in capturing all three distinctive features surrounding complexity [66].

In the context of technological entrepreneurship, for example, as maintained by McKelvey ([67], p. 67), "An entrepreneur could have co-evolutionary dynamics going on in his/her firm; a change in one part of a product leads to a change in another part, which then leads to further change in the part showing the initial change; these changes could affect marketing, production, supply chains, and so on. Finally, it could happen that an entirely new product appears. For example, think of all of the coevolving changes in computer, cell-phone, battery, and touch-screen technologies, computer programming, cell towers, the Internet, and the development of apps that led to current smart-phone products."

Similarly, in the context of technological ecosystems, Phillips and Ritala [68] interestingly build (and apply) a specific complexity-based, co-evolutionary framework. In particular, they suggest that three intertwined dimensions, i.e., conceptual (boundary and perspectives), structural (hierarchies and relationships), and temporal (dynamics and co-evolution) should be taken into account to understand (and predict) the behavior of complex ecosystems, especially in the case of an innovation (e.g., NPD) context.

Relatedly (and finally), as far as understanding the institutional complexity [69] of co-evolutionary ecosystems is specifically concerned, we are also in line with those scholars [70] who have recently claimed the increasing adoption of a neo-configurational perspective based on qualitative comparative analysis (QCA). Hence, for example, Misangyi [71] recently offered remarkable evidence regarding 28 business facilities projecting and implementing an environmental management system.

More generally, the claim towards the use of QCA is also in line with our claim above (please see Section 4.2.) that more qualitative research methodologies should be adopted to understand the complex nature of innovation-based settings. In this regard, for example, in a novel case study regarding innovation and change in organizational culture, Schlaile et al. [72] used a meme-based approach [73] to investigate the complexity-based interdependencies occurring in a German automotive consultancy firm.

5. Limitations and Conclusions

Through the results (and proposed implications) of this conceptual article, we do not aim to propose CT as the solution to all of the current EM sustainability-related issues. We also agree with those scholars who, seminally [74,75] or more recently [76,77], have identified the risks of transforming CT, when (even more generally) applied to management, as the fad of modern times. Specifically, we do not believe that this fast-growing approach will totally overwrite all of those theories based on positivism and reductionism [10,78].

Relatedly, we are also conscious that, from a methodological point of view, the results from our analysis present some limits, in that they are, at present, strictly focused on the leading journal in the EM field and on a static explanation. At this stage, in other words, our analysis of the 38 articles should be considered through the lens of a (hopefully useful) initial qualitative assessment, rather than the lens of a quantitative research, which has statistics and trends also aimed at being predictive. In this regard, however, we believe that our results could serve as a heuristic proxy, i.e., a conceptual start to be expanded through more journal-based searches and/or dynamic analyses.

In sum, although aware of the limitations above, and through discussing the implications of our findings, we attempted to explain how CT can contribute to govern many current issues associated with the EM research (concerning the research in, and practice of, sustainability issues). If firms are modeled as CAS, through the identification of agents, their interactions, feedback, and emergent phenomena, CT can then help find novel ways of working to foster a supposed desired emergent behavior (e.g., improved efficiency and effectiveness in NPD, team organization, technology management, or PM); thus, providing engineers and managers with new tools for improving decision-making and performance [79–81]. In this regard, for example, Bianchi et al. [82] innovatively deal with complexity management in a recent NPD context through a study of the interaction between stage-gate and agile models (and their associated principles to reduce uncertainty).

Of course, scholars and practitioners argue that, in order to be more than a metaphorical device, a relevant CT framework will need to always be more rigorous from the theoretical, mathematical, and computational modeling points of view [83,84]. We also believe that this modeling will need to be tested in different industry settings to ensure appropriate comparisons between models and real world structures [85–87]. In this way, CT may also be taken as a useful approach, for engineers and managers, to test the reliability and consistency of more conventional methods intended to improve sustainability.

In conclusion, firms, clusters, networks, and industries, may be seen, from some aspects, as similar to living organisms [88,89], which grow, evolve, and die [90,91]. They can be healthy or sick [92–94] and their behavior emerges from their internal qualities and dynamics, which provide complexity to the system, and from their interactions with the environment [95–97]. A firm's behavior is both affected by linear control, such as that imposed by bureaucracy or top-down management decisions, and natural, uncontrolled dynamics. If enterprise complexity fits the complexity of the environment, then desired behaviors, such as high performance and synergy, emerge [98,99].

To date, complexity represents one of the main problems surrounding sustainable business. While we think that the application of CT to business cannot eliminate this problem, we believe that it can help reduce it to a satisfying level.

Author Contributions: Conceptualization, G.A. and A.S.; methodology, G.A. and A.S.; investigation, A.S.; writing—original draft preparation, G.A. and A.S.; writing—review and editing, G.A. and A.S.; visualization, G.A. and A.S.; supervision, G.A. All authors have read and agreed to the published version of the manuscript.

Funding: This research received no external funding.

Conflicts of Interest: The authors declare no conflict of interest.

References

1. Philbin, S. Insights from managing complex research, technology and engineering projects in academia. *Eng. Manag. Syst. Eng. Videos* **2015**, *72*. Available online: https://scholarsmine.mst.edu/engman_syseng_videos/72 (accessed on 18 November 2020).
2. Daim, T. Editorial: The decade of technology intelligence. *IEEE Trans. Eng. Manag.* **2020**, *67*, 2–3. [CrossRef]
3. Ndofor, H.A.; Fabian, F.; Michel, J.G. Chaos in industry environments. *IEEE Trans. Eng. Manag.* **2018**, *65*, 191–203. [CrossRef]
4. McCarthy, I.P.; Tsinopoulos, C.; Allen, P.M.; Rose-Anderssen, C. New product development as a complex adaptive system of decisions. *J. Prod. Innov. Manag.* **2006**, *23*, 437–456. [CrossRef]
5. Amaral, L.A.N.; Uzzi, B. Complex systems-A new paradigm for the integrative study of management, physical, and technological systems. *Manag. Sci.* **2007**, *53*, 1033–1035. [CrossRef]
6. Baumann, O.; Siggelkow, N. Dealing with complexity: Integrated vs. chunky search processes. *Organ. Sci.* **2013**, *24*, 116–132.
7. Nicolis, G.; Prigogine, I. *Exploring Complexity: An Introduction*; W.H. Freeman: New York, NY, USA, 1989.
8. Waldrop, M.M. *Complexity: The Emerging Science at the Edge of Order and Chaos*; Simon & Schuster: New York, NY, USA, 1993.
9. Sibani, P.; Jensen, H.J. *Stochastic Dynamics of Complex Systems: From Glasses to Evolution*; Imperial College Press: London, UK, 2013.
10. Tsoukas, H. Don't simplify, complexify: From disjunctive to conjunctive theorizing in organization and management studies. *J. Manag. Stud.* **2017**, *54*, 132–153. [CrossRef]
11. Holland, J.H. *Complexity: A very Short Introduction*; Oxford University Press: Oxford, UK, 2014.
12. Price, I. Complexity, complicatedness and complexity: A new science behind organizational intervention? *Emergence* **2004**, *6*, 40–48.
13. Maguire, S.; McKelvey, B. Complexity and management: Moving from fad to firm foundations. *Emergence* **1999**, *1*, 19–61. [CrossRef]
14. Mitchell, M. *Complexity: A Guided Tour*; Oxford University Press: Oxford, UK, 2009.
15. Chartered Association of Business Schools. *Academic Journal Guide 2018*. 2018. Available online: https://charteredabs.org/academic-journal-guide-2018/ (accessed on 3 September 2020).
16. Mitleton-Kelly, E. *Complex Systems and Evolutionary Perspectives on Organisations: The Application of Complexity Theory to Organisations*; Elsevier Science: Oxford, UK, 2003.
17. Von Bertalanffy, L. *General System Theory. Development, Applications*; George Braziller: New York, NY, USA, 1968.
18. Holland, J.H. Emergence. *Philosophica* **1997**, *59*, 11–40.
19. Chiles, T.H.; Meyer, A.D.; Hench, T.J. Organizational emergence: The origin and transformation of Branson, Missouri's musical theaters. *Organ. Sci.* **2004**, *15*, 499–519. [CrossRef]
20. Parsons, T. *The Social System*; The Free Press: New York, NY, USA, 1951.
21. Beer, S. *Brain of the Firm*; The Penguin Press: London, UK, 1972.
22. Maturana, H.R.; Varela, F.J. *Autopoiesis and Cognition: The Realization of the Living*; D. Reidel: Boston, MA, USA, 1980.
23. Kauffman, S.A. *The Origins of Order: Self-Organization and Selection in Evolution*; Oxford University Press: New York, NY, USA, 1993.

24. Dominici, G.; Roblek, V.; Lombardi, R. A holistic approach to comprehending the complexity of the post-growth era: The emerging profile. In *Chaos, Complexity and Leadership 2014*; Erçetin, Ş., Ed.; Springer: Berlin/Heidelberg, Germany, 2016; pp. 29–42.
25. Carrubbo, L.; Iandolo, F.; Pitardi, V.; Calabrese, M. The viable decision maker for CAS survival: How to change and adapt through fitting process. *J. Serv. Theory Pract.* **2017**, *27*, 1006–1023. [CrossRef]
26. Secchi, D.; Neumann, M. (Eds.) *Agent-Based Simulation of Organizational Behavior*; Springer: Berlin/Heidelberg, Germany, 2016.
27. Pyka, A.; Grebel, T. Agent-based modelling–A methodology for the analysis of qualitative development processes. In *Agent-Based Computational Modelling*; Billari, F.C., Fent, T., Prskawetz, A., Scheffran, J., Eds.; Springer: Berlin/Heidelberg, Germany, 2006; pp. 17–35.
28. Fiss, P.C. Building better causal theories: A fuzzy set approach to typologies in organization research. *Acad. Manag. J.* **2011**, *54*, 393–420. [CrossRef]
29. Alyamani, R.; Long, S. The application of fuzzy Analytic Hierarchy Process in sustainable project selection. *Sustainability* **2020**, *12*, 8314. [CrossRef]
30. Tsilipanos, K.; Neokosmidis, I.; Varoutas, D. Modeling complex telecom investments: A system of systems approach. *IEEE Trans. Eng. Manag.* **2015**, *62*, 631–642. [CrossRef]
31. Barile, S.; Saviano, M. Complexity and sustainability in management. Insights from a systems perspective. In *Social Dynamics in a Systems Perspective*; Barile, S., Pellicano, M., Polese, F., Eds.; Springer: Berlin/Heidelberg, Germany, 2018; pp. 39–63.
32. Tranfield, D.; Denyer, D.; Smart, P. Towards a methodology for developing evidence-informed management knowledge by means of systematic review. *Br. J. Manag.* **2003**, *14*, 207–222. [CrossRef]
33. Polonioli, A. In search of better science: On the epistemic costs of systematic reviews and the need for a pluralistic stance to literature search. *Scientometrics* **2020**, *122*, 1267–1274. [CrossRef]
34. Breslin, D.; Gatrell, C. Theorizing through literature reviews: The miner-prospector continuum. *Organ. Res. Methods* **2020**. [CrossRef]
35. Newbert, S.L. Empirical research on the resource-based view of the firm: An assessment and suggestions for future research. *Strateg. Manag. J.* **2007**, *28*, 121–146. [CrossRef]
36. Phelps, R.; Adams, R.; Bessant, J. Life cycles of growing organizations: A review with implications for knowledge and learning. *Int. J. Manag. Rev.* **2007**, *9*, 1–30. [CrossRef]
37. Poggesi, S.; Mari, M.; De Vita, L.; Foss, L. Women entrepreneurship in STEM fields: Literature review and future research avenues. *Int. Entrep. Manag. J.* **2020**, *16*, 17–41. [CrossRef]
38. Xirogiannis, G.; Glykas, M. Fuzzy cognitive maps in business analysis and performance-driven change. *IEEE Trans. Eng. Manag.* **2004**, *51*, 334–351. [CrossRef]
39. Shafiei-Monfared, S.; Jenab, K. Fuzzy complexity model for enterprise maintenance projects. *IEEE Trans. Eng. Manag.* **2012**, *59*, 293–298. [CrossRef]
40. Lin, C.T.; Chen, C.T. New product go/No-go evaluation at the front end: A fuzzy linguistic approach. *IEEE Trans. Eng. Manag.* **2004**, *51*, 197–207. [CrossRef]
41. Mikaelian, T.; Nightingale, D.J.; Rhodes, D.H.; Hastings, D.E. Real options in enterprise architecture: A holistic mapping of mechanisms and types for uncertainty management. *IEEE Trans. Eng. Manag.* **2011**, *58*, 457–470. [CrossRef]
42. Giannoccaro, I.; Nair, A. Examining the roles of product complexity and manager behavior on product design decisions: An agent-based study using NK simulation. *IEEE Trans. Eng. Manag.* **2016**, *63*, 237–247. [CrossRef]
43. Tripathy, A.; Eppinger, S.D. Organizing global product development for complex engineered systems. *IEEE Trans. Eng. Manag.* **2011**, *58*, 510–529.
44. Levardy, V.; Browning, T.R. An adaptive process model to support product development project management. *IEEE Trans. Eng. Manag.* **2009**, *56*, 600–620. [CrossRef]
45. Jun, H.B.; Suh, H.W. A modeling framework for product development process considering its characteristics. *IEEE Trans. Eng. Manag.* **2008**, *55*, 103–119. [CrossRef]
46. Pathak, S.D.; Dilts, D.M.; Biswas, G. On the evolutionary dynamics of supply network topologies. *IEEE Trans. Eng. Manag.* **2007**, *54*, 662–672.
47. Marzi, G.; Caputo, A.; Garces, E.; Dabic, M. A three decade mixed-method bibliometric investigation of the IEEE Transactions on Engineering Management. *IEEE Trans. Eng. Manag.* **2020**, *67*, 4–17. [CrossRef]

48. Uhl-Bien, M.; Marion, R.; McKelvey, B. Complexity leadership theory: Shifting leadership from the industrial age to the knowledge era. *Leadersh. Q.* **2007**, *18*, 298–318. [CrossRef]
49. Burnes, B. Complexity theories and organizational change. *Int. J. Manag. Rev.* **2005**, *7*, 73–90. [CrossRef]
50. Surace, A. Complexity and leadership: The case of a military organization. *Int. J. Organ. Anal.* **2019**, *27*, 1522–1541. [CrossRef]
51. Emery, F.; Trist, E. The causal texture of organizational environments. *Hum. Relat.* **1965**, *18*, 21–32. [CrossRef]
52. Lee, B.; Saunders, M. *Conducting Case Study Research for Business and Management Students*; Sage: London, UK, 2017.
53. Brown, S.L.; Eisenhardt, K.M. The art of continuous change: Linking complexity theory and time-paced evolution in relentlessly shifting organizations. *Adm. Sci. Q.* **1997**, *42*, 1–34. [CrossRef]
54. Eisenhardt, K.M.; Bhatia, M.M. Organizational complexity and computation. In *Companion to Organizations*; Baum, J.A.C., Ed.; Blackwell: Oxford, UK, 2002; pp. 442–466.
55. Jacobs, M.A. Complexity: Toward an empirical measure. *Technovation* **2013**, *33*, 111–118. [CrossRef]
56. Chiva, R.; Ghauri, P.; Alegre, J. Organizational learning, innovation and internationalization: A complex system model. *Br. J. Manag.* **2014**, *25*, 687–705. [CrossRef]
57. Anderson, P. Complexity theory and organization science. *Organ. Sci.* **1999**, *10*, 216–232. [CrossRef]
58. McKelvey, B. Avoiding complexity catastrophe in co-evolutionary pockets: Strategies for rugged landscapes. *Organ. Sci.* **1999**, *10*, 294–321. [CrossRef]
59. Cafferata, R. Darwinist connections between the systemness of social organizations and their evolution. *J. Manag. Gov.* **2016**, *20*, 19–44. [CrossRef]
60. Sandhu, S.; Kulik, C. Shaping and being shaped: How organizational structure and managerial discretion co-evolve in new managerial roles. *Adm. Sci. Q.* **2019**, *64*, 619–658. [CrossRef]
61. Abatecola, G.; Breslin, D.; Kask, J. Do organizations really co-evolve? Problematizing co-evolutionary change in management and organization studies. *Technol. Forecast. Soc. Chang.* **2020**, *155*, 119964. [CrossRef]
62. Paniccia, P.M.A.; Baiocco, S. Interpreting sustainable agritourism through co-evolution of social organizations. *J. Sustain. Tour.* **2021**, *29*, 87–105. [CrossRef]
63. Volberda, H.; Lewin, A. Co-evolutionary dynamics within and between firms: From evolution to co-evolution. *J. Manag. Stud.* **2003**, *40*, 2111–2136. [CrossRef]
64. Murmann, J.P. The co-evolution of industries and important features of their environments. *Organ. Sci.* **2013**, *24*, 58–78. [CrossRef]
65. Almudi, I.; Fatas-Villafranca, F.; Izquierdo, L.R.; Potts, J. The economics of Utopia: A co-evolutionary model of ideas, citizenship and socio-political change. *J. Evol. Econ.* **2017**, *27*, 629–662. [CrossRef]
66. Adinolfi, P. A journey around decision-making: Searching for the "big picture" across disciplines. *Eur. Manag. J.* **2020**. [CrossRef]
67. McKelvey, B. Complexity ingredients required for entrepreneurial success. *Entrep. Res. J.* **2016**, *6*, 53–73. [CrossRef]
68. Phillips, M.A.; Ritala, P. A complex adaptive systems agenda for ecosystem research methodology. *Technol. Forecast. Soc. Chang.* **2019**, *148*, 119739.
69. Greenwood, R.; Raynard, M.; Kodeih, F.; Micelotta, E.R.; Lounsbury, M. Institutional complexity and organizational responses. *Acad. Manag. Ann.* **2011**, *5*, 317–371. [CrossRef]
70. Misangyi, V.F.; Greckhamer, T.; Furnari, S.; Fiss, P.C.; Crilly, D.; Aguilera, R. Embracing causal complexity: The emergence of neo-configurational perspective. *J. Manag.* **2017**, *43*, 255–282. [CrossRef]
71. Misangyi, V.F. Institutional complexity and the meaning of loose coupling: Connecting institutional sayings and (not) doings. *Strateg. Organ.* **2016**, *14*, 407–440. [CrossRef]
72. Schlaile, M.; Bogner, K.; Muelder, L. It's more than complicated! Using organizational memetics to capture the complexity of organizational culture. *J. Bus. Res.* **2019**. [CrossRef]
73. Price, I. Organizational memetics? Organizational learning as a selection process. *Manag. Learn.* **1995**, *26*, 299–318. [CrossRef]
74. Simon, H.A. The architecture of complexity. *Proc. Am. Philos. Soc.* **1962**, *106*, 467–482.
75. Perrow, C. *Complex Organizations: A Critical Essay*; Scott Foresman: Glenview, IL, USA, 1972.
76. McKelvey, B. Quasi-natural organization science. *Organ. Sci.* **1997**, *8*, 352–380. [CrossRef]
77. Lissack, M.R.; Gunz, H.P. (Eds.) *Managing Complexity in Organizations: A View in Many Directions*; Quorum Books: Westport, CT, USA, 1999.

78. McKelvey, B. (Ed.) *Complexity: Critical Concepts*; Routledge: Oxford, UK, 2013.
79. Allen, P.M.; Maguire, S.; McKelvey, B. (Eds.) *The SAGE Handbook of Complexity and Management*; Sage: London, UK, 2011.
80. Boulton, J.G.; Allen, P.M.; Bowman, C. *Embracing Complexity: Strategic Perspectives for an Age of Turbulence*; Oxford University Press: Oxford, UK, 2015.
81. Cristofaro, M. Reducing biases of decision-making processes in complex organizations. *Manag. Res. Rev.* **2017**, *40*, 270–291.
82. Bianchi, M.; Marzi, G.; Guerini, M. Agile, stage-gate and their combination: Exploring how they relate to performance in software development. *J. Bus. Res.* **2020**, *110*, 538–553. [CrossRef]
83. Ofori-Dankwa, J.; Julian, S.D. Complexifying organizational theory: Illustrations using time research. *Acad. Manag. Rev.* **2001**, *26*, 415–430. [CrossRef]
84. Andriani, P.; McKelvey, B. From Gaussian to Paretian thinking: Causes and implications of power laws in organizations. *Organ. Sci.* **2009**, *20*, 1053–1071. [CrossRef]
85. Linn, S.; Tay, N.S.P. Complexity and the character of stock returns: Empirical evidence and a model of asset prices based on investor learning. *Manag. Sci.* **2007**, *53*, 1165–1181. [CrossRef]
86. Wu, P.L.; Yeh, S.S.; Woodside, A.G. Applying complexity theory to deepen service dominant logic: Configural analysis of customer experience-and-outcome assessments of professional services for personal transformations. *J. Bus. Res.* **2014**, *67*, 1647–1670. [CrossRef]
87. Padalkar, M.; Gopinath, S. Are complexity and uncertainty distinct concepts in project management? A taxonomical examination from literature. *Int. J. Proj. Manag.* **2016**, *34*, 688–700. [CrossRef]
88. Belussi, F.; Sedita, S.R. Industrial districts as open learning systems: Combining emergent and deliberate knowledge structures. *Reg. Stud.* **2012**, *46*, 165–184. [CrossRef]
89. Ferraro, G.; Iovanella, A. Organizing collaboration in inter-organizational innovation networks, from orchestration to choreography. *Int. J. Eng. Bus. Manag.* **2015**, *7*, 7–24. [CrossRef]
90. Hodgson, G.; Knudsen, T. *Darwin's Conjecture: The Search for General Principles of Social and Economic Evolution*; University of Chicago Press: Chicago, IL, USA, 2010.
91. Grandinetti, R. Is organizational evolution Darwinian and/or Lamarckian? *Int. J. Organ. Anal.* **2018**, *26*, 858–874. [CrossRef]
92. MacIntosh, R.; MacLean, D.; Burns, H. Health in organization: Towards a process-based view. *J. Manag. Stud.* **2007**, *44*, 206–221. [CrossRef]
93. Ciampi, F.; Gordini, N. Small enterprise default prediction modeling through artificial neural networks: An empirical analysis of Italian small enterprises. *J. Small Bus. Manag.* **2013**, *51*, 23–45. [CrossRef]
94. Hristov, I.; Chirico, A.; Appolloni, A. Sustainability value creation, survival, and growth of the company: A critical perspective in the Sustainability Balanced Scorecard (SBSC). *Sustainability* **2019**, *11*, 2119. [CrossRef]
95. Jones, C. An autecological interpretation of the firm and its environment. *J. Manag. Gov.* **2016**, *20*, 69–87. [CrossRef]
96. Mingione, M.; Leoni, L. Blurring B2C and B2B boundaries: Corporate brand value co-creation in B2B2C markets. *J. Mark. Manag.* **2020**, *36*, 72–99. [CrossRef]
97. Sarta, A.; Durand, R.; Vergne, J.P. Organizational adaptation. *J. Manag.* **2020**. [CrossRef]
98. Diaz-Fernandez, M.C.; Gonzalez-Rodriguez, M.R.; Simonetti, B. Top management team diversity and performance: An integrative approach based on Upper Echelons and complexity theory. *Eur. Manag. J.* **2020**, *38*, 157–168. [CrossRef]
99. Zachary, D.; Dobson, S. Urban development and complexity: Shannon entropy as a measure of diversity. *Plan. Pract. Res.* **2020**. [CrossRef]

Publisher's Note: MDPI stays neutral with regard to jurisdictional claims in published maps and institutional affiliations.

Article

Machine Learning Based Vehicle to Grid Strategy for Improving the Energy Performance of Public Buildings

Connor Scott, Mominul Ahsan * and Alhussein Albarbar

Smart Infrastructure and Industry Research Group, Department of Engineering, Manchester Metropolitan University, Chester St., Manchester M1 5GD, UK; connor.scott@stu.mmu.ac.uk (C.S.); a.albarbar@mmu.ac.uk (A.A.)
* Correspondence: md.ahsan2@mail.dcu.ie

Abstract: Carbon neutral buildings are dependent on effective energy management systems and harvesting energy from unpredictable renewable sources. One strategy is to utilise the capacity from electric vehicles, while renewables are not available according to demand. Vehicle to grid (V2G) technology can only be expanded if there is funding and realisation that it works, so investment must be in place first, with charging stations and with the electric vehicles to begin with. The installer of the charging stations will achieve the financial benefit or have an incentive and vice versa for the owners of the electric vehicles. The paper presents an effective V2G strategy that was developed and implemented for an operational university campus. A machine learning algorithm has also been derived to predict energy consumption and energy costs for the investigated building. The accuracy of the developed algorithm in predicting energy consumption was found to be between 94% and 96%, with an average of less than 5% error in costs predictions. The achieved results show that energy consumption savings are in the range of 35%, with the potentials to achieve about 65% if the strategy was applied at all times. This has demonstrated the effectiveness of the machine learning algorithm in carbon print reductions.

Keywords: carbon neutral; electric vehicle; vehicle-to-grid; renewable energy; smart charging; net-zero

Citation: Scott, C.; Ahsan, M.; Albarbar, A. Machine Learning Based Vehicle to Grid Strategy for Improving the Energy Performance of Public Buildings. *Sustainability* **2021**, *13*, 4003. https://doi.org/10.3390/su13074003

Academic Editor: Simon Philbin

Received: 7 March 2021
Accepted: 1 April 2021
Published: 3 April 2021

1. Introduction

The future of power generation will be dominated by harvesting energies from renewable sources and methods for reducing CO_2 emissions. This is ensured through government regulations with, e.g., the UK's net-zero greenhouse gas (GHG) target by 2050 [1]. As the demand for energy increases, the reliability of the generation and distribution network of national grids (NGs) decreases, due to intermittent renewable energy sources. This can be unpredictable, as the supply can be low when the renewable generation is high, and the volatility of renewable generation can create an unpredictable demand from the NG. In France, the decrease in fossil fuels leads to an increase in renewables, mostly wind, which leads to the greater import of energy and a decrease in grid consistency [2].

To compensate for the inefficiency of renewable power supply installations, Vehicle to grid (V2G) can be used. In the V2G method, the batteries from charged electric vehicles (EV) can be discharged back into the grid at peak times. It can be connected to the grid, or it can be utilised by a business, which can choose to store the energy, sell to the national grid, or use it. Peak power plants are used when the demand exceeds what is expected [3]. These plants can quickly generate the required energy on top of the already generated energy of the NG, but they are not environmentally friendly, with the fuel being sourced by gas or diesel generating CO_2 as a bio-product. Therefore, it is required to develop a way of excess peak power generation without the use of fossil fuels.

1.1. UK EVs and Power Generation

In March 2019, it was estimated 38.4 million licensed vehicles in the UK [4] with 317,000 ultra-low emission vehicles (ULEV) recorded [5], and only 25,000 recorded EV charging points, mostly located in dense areas, such as cities. The EV market should scale 100% of new sales by 2030, making it a future leader in-vehicle use in the UK. As the number of EVs increase, the infrastructure and techniques used for charging-discharging will have to be adapted and improved. If there is a lack of EV chargers compared to the number of EVs, the EV market will not grow because many EV owners will not be able to charge, and use, their car. The UK's demand is 38.58 GW at 10:30 on 4 February 2020 [6], with an average EV battery size being 37.125 kW [7], meaning the EV battery capacity in the UK is 26.3% of the total demand. The sizeable, combined EV battery capacity can reduce stress and increase profitability for both the NG and businesses that employ V2G. However, the low number of charging points means this is reduced to 2.4%, which is a big loss. More EV charging points will enable more V2G to take place, meaning the EV battery capacity can be utilised more for the benefit of the grid and the business that employs it. The amount of EV charging points will obviously increase with the rising EVs. Renewable production equates to 31.49% wind, 2.36% hydro and 4.71% solar, which is 38.56% of the production. Wind power varies between 0.25–12.5 GW, hydro 0–0.5 GW and solar varies between 0–4 GW per week. This equates to a variability of 16.75 GW per week, which is 43% of the demand. The EVs' current capacity is more than half of the renewable variation.

1.2. EV Infrastructure and Net Profits

The success of the V2G solution relies on consumers' cooperation, and thus, it is dictated by consumers' perspective. Location of charging stations play a great role in using EVs as some people prefer to charge their EV at home, whereas some of them do at work. The number of charging stations play a role as well, as there may not be available enough at their place of work either. This creates a disadvantage with the range of the vehicle, as using conventional fuelled vehicles allow long-distance travel and quick fuelling, whereas EV's are limited by distance and charging time. Some areas are situated without charging points for 10's of miles, thus, the availability of charging points is essential for future use of EVs.

According to 'Office for Low Emission Vehicles' (OLEV), an EV charger will cost roughly £650 to purchase and install with a Government incentive to pay up to £500 of the cost [8]. This would eliminate the problem of people without having enough charging points as they could charge them at home. While charging at home, people may not charge their EV's at work, in the denser areas where bi-directional EV charging is essential. UK Government offers a grant towards EVs as they are more costly than ICE's. This grant could be up to 35% on cars passing certain requirements and 20% on motorcycles, mopeds, vans and taxis of up to £3500, £1500, £1500, £8000 and £7500, respectively.

The importance of the financial side of the method is discussed through investigating the costs and benefits of EVs participating in a V2G scheme. The work in Reference [9] uses four brands of EV, showing the trend between energy use from the battery and the higher price for peak-time, which can generate income. As the battery is being charged and discharged more commonly, the lifespan may drastically decrease where the consumers require buying new batteries more often, so the cheaper the battery, the higher the net income. The highest cost of V2G investment by power grid companies in the battery storage needed. The larger the battery capacity of EVs, the fewer EVs are needed to match the same capacity, meaning the reduction of investment costs for battery storage from the NG.

1.3. Lifespan and Variables of EVs

An average motorists' mileage is around 7600 miles per year, and the battery capacity for an EV may decrease to 80% after 20 years of driving [10,11]. A certain model of EV was driven with a mileage of 232,442 miles and had a range of 220 miles of its original 265-mile range. This equates to 83% of the original battery life, meaning the batteries for an EV should last at least 20 years or even longer [10]. Another EV model estimated battery life to be 10–12 years beyond the life of the car, with Nissan's evaluation of a 10-year car life and up to 22 years for the lifespan of the battery [12].

The EV is usually under perfect driving conditions when tested or simulated, whereas the driving range can be more accurately predicted using various methods. A 'Fuzzy Logic Classifier' method was used for conducting simulation and prediction. Certain variables are necessary to get a prediction, such as the battery size and weights of 1, 2 and 4 people in the car. This also includes variables, such as the slope, in which the vehicle is travelling to calculates a more accurate prediction on the battery lifespan [13]. EV batteries are continuing to improve, with different companies improving at different rates. The lifespan of the battery plays a great role in the proposed method, as the owner of the EV is compensated for the reduction in battery lifespan. The range on a type of EV battery is reduced by 2% for every 100,000 km [14]. As batteries become more efficient and last longer, this does not make as much of an impact on the method.

1.4. Methods of V2G Charging

Several techniques are employed for charging and discharging EVs [15]. Controlled charge-discharge allows the operator to decide when EV will be charged and discharged, giving more freedom and control to the grid. If the EV is plugged in at peak demand, the operator can hold charging of the vehicle until the demand is less. Uncontrolled charge-discharge starts to charge the EV as soon as it is plugged in. It may have a great effect on the reliability of the NG, producing problems, including a greater variation of frequency, demand, and overall, reliability. Therefore, this method is unlikely to find its way into the future of EV charging. Intelligent charging uses real-time energy demand to decide whether the EVs will be charged or discharged, using grid requirements, allowing the operator to maintain the quality of the grid, such as the frequency and voltage etc. Indirect controlled charging works from the users' perspective, allowing the user to charge their car for a lower price or possibly in the future for a profit, as the EVs can be charged at off-peak and then discharged at peak times, generating a small profit.

1.5. Previous V2G Simulations

The V2G method has been proposed for use in domestic applications, but unfortunately, without the incentives for the EV owners [16,17]. These simulations showed the volatility of the system and the difficulty in predicting the future supply and demand, meaning the simulations can be inaccurate. The supplier of the energy, in this case, has great advantages, through having a less volatile demand, as the V2G method provides peak shaving. Attention should be given to the V2G method for predicting energy demand rather than financial benefits for both parties. Various topologies of V2G with dependence existed where the EVs are connected to the grid. However, 'Vehicle to building' (V2B), 'Vehicle to home' (V2H) and 'Vehicle to load' (V2L) are the same methods, which covers all variations in the process [18].

The depth of discharge is a contributing factor to the results within this method, as the battery that is discharged less will have a longer lifespan [19]. To improve the lifespan of the battery, if the battery does not need to be discharged, then it shouldn't be. Fuel cells can be combined with the V2G technology to provide a more efficient method. With the application of fuel cell vehicles (FCV), there can be a 51% increased income as the required supply is spread out over FCV and V2G instead of using only one method [20].

As the total cost of ownership (TCO) of EVs can vary, changing consumers' opinion on if they will use one or not. This is being amended by the tariff schemes given by the government [21]. Xcel energy's off-peak EV rate is 4.3 cents per kWh compared to on-peak rates of 17 cents, with a four-fold increase, the time of purchase creates an incentive. A suitable EV infrastructure, and smart EV chargers, which enable the demands of the consumer to be met, will allow the expansion of EVs. Carbon neutral building will integrate the use of V2G systems and EVs, whether it's V2G or just the use of EVs rather than standard vehicles. It is undeniable that the use of EVs reduces the carbon footprint of the building on a university campus [22]. EVs and hybrid electric vehicles (HEVs) are becoming more efficient and useable. As the V2G method requires the use of EVs, further research into better performance and attractiveness to consumers' increases the value of the V2G method [23].

V2G methods are not unfamiliar, as is shown in previous research papers. The application of V2G methods into large public and commercial buildings, e.g., university campus, validated by data collected from operational environments and critically analysed by employing advanced algorithms, have not been investigated yet. The surrounding aspects, such as the vehicle range, using ML have been previously analysed without focusing on the buildings and EV users' financial and environmental benefits.

A V2G system is currently being tested in the UK, with no mention of how the system is going to be set up or analysed. There is not enough information available in the literature for predicting energy management of public buildings, particularly with employing machine learning (ML) algorithms [24]. A machine learning algorithm can enhance the quality of projects, such as these, by giving the company an accurate outcome, regarding the financial and energy characteristics of the project. The 'Parkers Vehicle-Grid Integration Summit' showed the implementation of the method, and then the obtaining of the results [25]. The method proposed in Section 2 can predict the outcome before the implementation.

The problem surrounding the growth of EVs and V2G systems is uncertain for both the driver and the building. It is not only difficult to predict the existing V2G system, but also the future of the V2G system needs to be predicted considering the location, scale of the building(s), future improved efficiency in EV batteries, etc.

In this paper, an effective V2G scheme has been developed and implemented into an operational university building, enabling it to be used on a larger scale. Parameters and system requirements, considering the initial cost, and net profit for both the installers and users of the EV chargers for the campus, are shown in Section 2. The on-site energy storage's price is calculated and equated to EV chargers, cost of EV charging stations, their lifespan and the campus profit have been presented in Section 3. In addition, the net profit of both parties, considering the long-term effects of using V2G, application of machine learning (ML) for predicting energy consumption and cost have also been conducted in this section. Critical analysis of achieved results with comparisons has also been presented in this section. Finally, key conclusions are drawn, and future works are suggested in Section 4.

2. Proposed Methodology

2.1. The Proposed Methodology

The purpose of V2G methods is to acquire energy at peak demand, while still meeting the target of net-zero GHG emissions. The V2G method presented in Figure 1 was designed and applied to an operational University campus. The method is determined by the tariff rate, which is determined by energy demand that depends on the time of day. While the energy demand is high, the tariff rate is high, and the energy can be taken from the EVs instead of buying at a higher price from the electricity grid. When the energy demand is low, the EVs can be charged, as the electricity can be bought from the grid. This method works under the proposition that the EVs will leave at 17:00, as the most common times of use is 09:00–17:00. The EV must be fully charged by 17:00, which is enabled by the above chart.

The English north-west university building used 2,666,560 kW in 2019, with a daily average of 7936.19 kW with the energy peaking at the university opening hours from 09:00 to 17:00. The university pays 12 p/kW off-peak and 15 p/kW on the peak. The average EV car battery size is 37.125 kW meaning each charger would provide 27.75 kW of battery storage for the campus. EV batteries can only be discharged to 25% to stop critical battery degradation [26]. This method uses the batteries from the EVs to supply the campus at peak times instead of buying at peak prices or having a large battery installed.

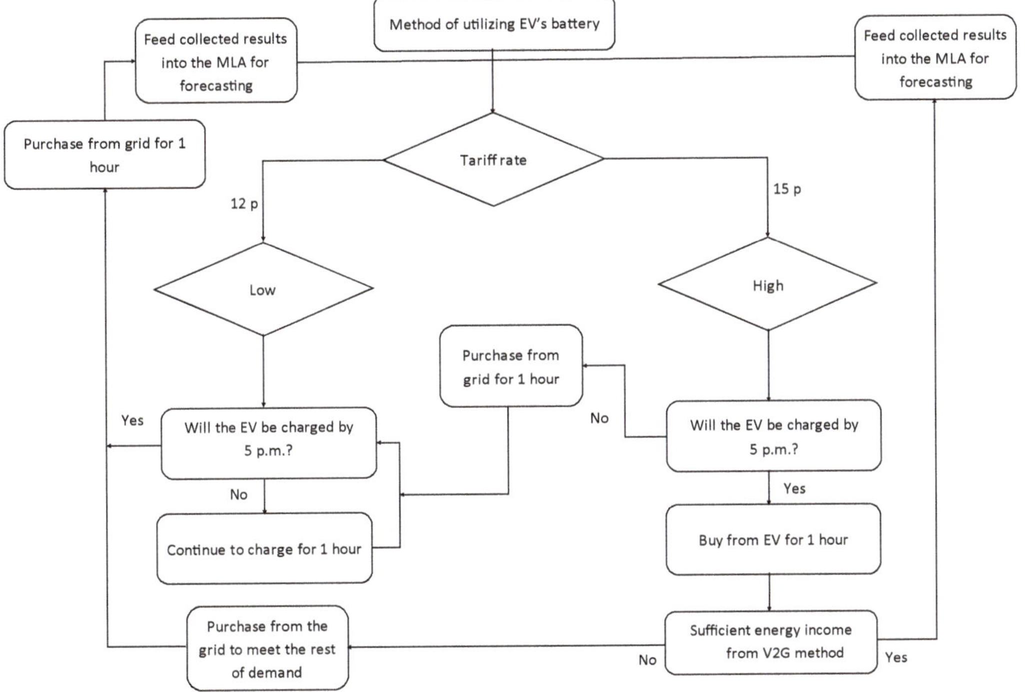

Figure 1. The methodological steps of the V2G implementation.

To calculate the costs and savings of charger installation, the company 'PodPoint' [27] was used. This was a quote given for purchasing and installing the charger. Equation (1) shows the profit (*P*) calculated using money made (*M*) and installation cost (*I*).

$$P = M - I \quad (1)$$

The period of return on investment (ROI) was calculated by plotting the 'Money made' through time until the profit is no longer a negative. Instead of buying electricity when the renewable harvesting is low, the electricity from the EV batteries can be used. This saves money as instead of buying expensive electricity from the grid at peak times, the EV batteries can be discharged and then recharged at off-peak times. In addition, the cost of large-scale batteries will outweigh the purchase and installation costs of charging stations.

The electricity usage of the investigated building (Business School Building—BSB) was 8256 kW. Where '*t*' is time, '*Ah*' is Amps per hour, '*w*' is Watts, '*Nb*' is the number of batteries and '*C*' is battery capacity, the storage for this can be calculated through Equations (2) and (3).

$$Ah = \frac{w}{t} \quad (2)$$

$$Nb = \frac{Ah}{C} \quad (3)$$

The cost and number of batteries can vary significantly depending on the required capacity. A battery of this size would be roughly £1823.86, which is too expensive to insure a good ROI. This would be costly for any ROI or to be beneficial to anyone involved, meaning the V2G method is essential. Instead of using an on-site battery, the EVs plugged into the charging points can be used as they're probably going to be there during peak times.

EV battery cost determines the net-profit of the car owner as the battery will need to be replaced more often as it is being charged and discharged more frequently. A cost analysis of the university campus and EV users was carried out to determine if this was successful for both parties, which has been further elaborated in the results section.

The proposed method is a novel algorithm that is capable of accurately predicting the present, and future, financial and energy characteristics of the vehicle-to-grid system. The collected energy characteristic data from the building will train the algorithm and improve accuracy and usability.

2.2. Machine Learning (ML) Algorithm

For a large amount of data, machine learning (ML), including supervised and unsupervised techniques, are massively used, especially for classification problems [28]. Input and output data are required to build and train the supervised model, which will be used in predicting future outcome for relevant new data sets. On the contrary, only input data are enough in developing models using unsupervised learning [29,30]. Machine learning was used for predicting energy savings for a building in Reference [31]. The methods used multiple linear regression, support vector regression, and back-propagation neural network. ML was also used for the thermal response of buildings, e.g., comparisons between measured and predicted results. The thermal load of a building was predicted using machine learning. The ML approach proved useful, depending on whether it was predicting short-term or long-term, and the predicted data was most accurate when supplied with uncertain weather data. The approach was used to predict only the thermal load of a building [32]. The behaviour of residential buildings has been predicted using machine learning techniques in Reference [33]. The model required a small training set while predicting accurately. The method use was the holt-winters extreme learning machine. This proved to be accurate while only needing 50 days of input data.

As shown in Figure 2, the neural networks (NN) based model starts with an input layer, through the weights, into the hidden layer, more weights, and into the output. The input is multiplied by the weights to reduce the error. Each hidden layer function is specialised to produce a defined output.

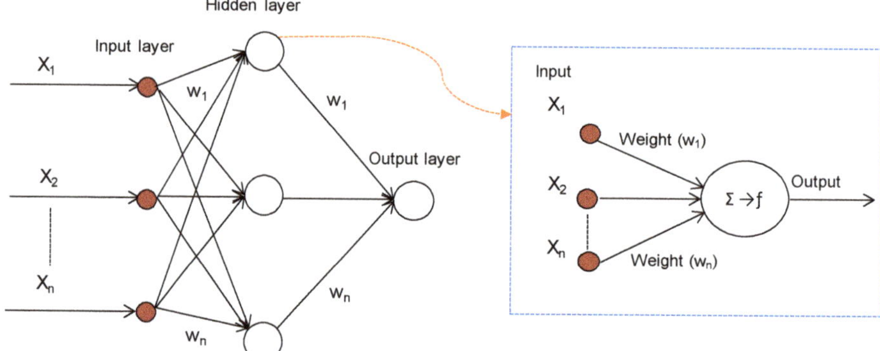

Figure 2. NN model structure.

It is required to arrange input, output, and validation data for developing a NN model and making a prediction. After creating the model employing input data, validation data set are used for conducting prediction of the new data through the developed model [34]. In NN, the gradient descent method and the Gauss–Newton method are quite popular. Particularly for solving nonlinear problems, the algorithms use its standard technique [35]. Mean squared error performs an evaluation of the training performance through simplifying the construction of a network by minimising the sum of the squared errors.

Hessian matrix approximates the sum of squares by $H = J^T J$, where J is a Jacobian matrix, gradient $g = J^T e$. e is denoted as network error and Levenberg-Marquardt training algorithm can be presented by Equation (4).

$$X_{k+1} = X_k - [J^T J + \mu I]^{-1} J^T e \quad (4)$$

Connection weight is called by 'X_k' at the 'k' number of iterations. Scalar combination coefficient is denoted by 'μ', which accomplishes transformation to either gradient descent or Gauss–Newton algorithm. The identity matrix is represented by 'I', where the training process of NN is performed through the descent gradient method as a learning rule. The error between the outcome of training and the targeted output is calculated through the error function. All the calculation is performed by determining the sum of the squared errors of input data and output patterns of the training set. The error function is calculated by Equation (5).

$$\varepsilon = \sum t_p - f_p \quad (5)$$

Where 't_p' indicates targeted output and 'f_p' denotes actual output. The goal of using this learning rule is to search for suitable values of weights for minimising the error.

A feed-forward neural network (FFNN) machine learning algorithm (MLA) was used to predict the future energy demand of the investigated building. The data taken from 2013–2017 was used as an input and built ML models where the years 2018 and 2019 were predicted, as depicted in Figure 3.

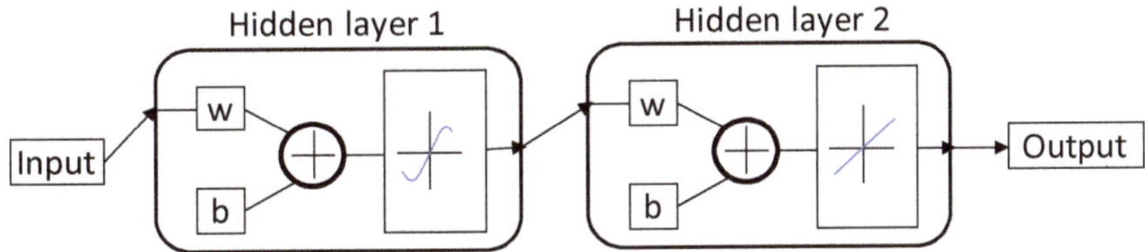

Figure 3. Architecture of neural network.

The proposed approach allows the training of data for the aim of building models and future prediction. In our study, the energy demand of the building was predicted for the years 2018 and 2019. In addition, the V2G cost was calculated using the energy demand for 2013–2017, and the cost was predicted for 2018 and 2019, which have been presented in the results section.

3. Results and Discussion

3.1. On-Site Battery Storage

Excess electricity is stored in a battery where the size of the battery is dependent on how much electricity is required to store. From the simulation, variables, such as the type of vehicle, battery capacity, charging rate, number of charging stations, kW rating of charging station and price of electricity, has been calculated. The ROI and overall profit for both the consumer and for the stand-alone building can be determined. The 8256 kW comes from the demand for the building on a given day—see Equations (6) and (7).

$$Ah = \frac{8256\ kW}{12} = 688\ kA \quad (6)$$

$$Nb = \frac{688\ kA}{4560\ A/h} = 150.87 = 151 \quad (7)$$

This can be converted to cover the capacity of 10 EVs connected to the campus in Equation (8).

$$SC = Cr \times Nc \times Nh = 6.66 \times 10 \times 12 = 799\ kW/d \tag{8}$$

SC is the charging station capacity, Cr is the charging rate of the station, Nc is the number of charging stations, Nh is the number of hours they will be needed for, and Nb is the number of batteries. The battery capacity for this can be calculated using Equations (9) and (10).

$$Ah = \frac{799.2\ kW}{12} = 66.6\ kA \tag{9}$$

$$Nb = \frac{66.6\ kA}{4560\ Aph} = 14.6 = 15\ Batteries \tag{10}$$

The price is estimated as £27,432.90 for 15 batteries, where the price of the battery can be affected by the factors, including capacity, size and the supplier.

3.2. Charging Stations, EV Variability and Campus Profit

The time of day, price of electricity and net profit for all involved in this V2G method are analysed and explained. If the campus buys at off-peak prices, not giving the EV owners an incentive, they make 43.56 p/day using a 3.63 kW/h charger (Table 1).

Table 1. Energy sale management based on time and price between EV owner and the university campus.

1. Time	2. Price (Pence)	3. Charge/ Discharge	4. EV Owner (Pence)	5. Campus (Pence)	6. Campus at Off-Peak Prices (Pence)
08:00–09:00	43.56	Charge	−43.56	+43.56	+43.56
09:00–10:00	54.45	Discharge	+54.45	−54.45	−43.56
10:00–11:00	54.45	Discharge	+54.45	−54.45	−43.56
11:00–12:00	43.56	Charge	−43.56	+43.56	+43.56
12:00–13:00	54.45	Discharge	+54.45	−54.45	−43.56
13:00–14:00	54.45	Discharge	+54.45	−54.45	−43.56
14:00–15:00	43.56	Charge	−43.56	+43.56	+43.56
15:00–16:00	43.56	Charge	−43.56	+43.56	+43.56
16:00–17:00	43.56	Charge	−43.56	+43.56	+43.56
		Net Profit	0	0	+43.56

Figure 4 demonstrates how the profit can vary throughout the day; if the peak-time, and off-peak prices are paid for the electricity, there will be no profit. The 'x' axis represents the hours of charge and the 'y' axis represents net profit in pence. There are more discharge hours than charge hours for the EVs, meaning the building must purchase the electricity at 15 p/h and must sell it back to the EV at 12 p/h to break even. The main operating hours are 09:00–17:00, with 12 p showing the peak times and −15 p showing off-peak times for the building for an average day.

A 3.63 kW/h charger's values were calculated using the EVs' batteries at peak times and re-charging at off-peak prices. It also shows that if the University charges the EV batteries at off-peak rates instead of peak rates, the EV owner's and the campus can profit, as is shown in columns 4 and 5 in Table 1. If the university buys and sells at off-peak prices, it makes money throughout the day. While the campus pays both peak, and off-peak prices with a 1 kW/h charger, it breaks even, as the last payment of the day takes net profit to zero (Figure 4).

The tariff rate changes throughout the day. The low tariff rate is 12 p/h, and the high tariff rate is 15 p/h. When these prices are paid, the net profit ends at zero at 17:00. If the campus pays off-peak prices to the EV owner for any hour of the day, 12 p/day is earned as the last payment of the day takes net profit to 12 p. Figure 5 presents the outcome of net profit.

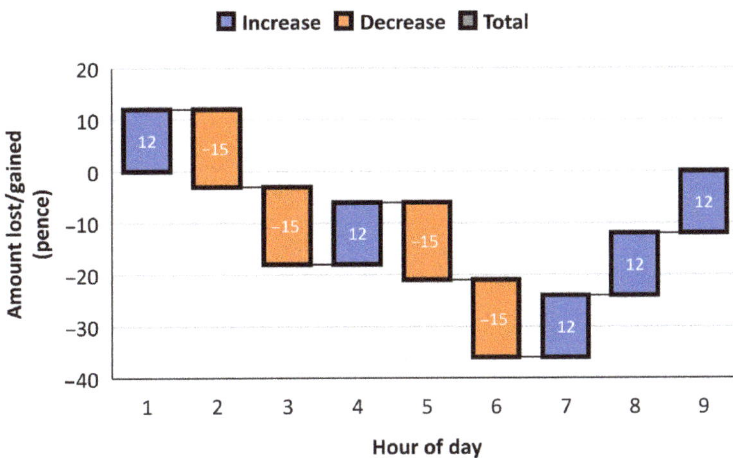

Figure 4. Campus net profit while peak-time and off-peak prices are paid.

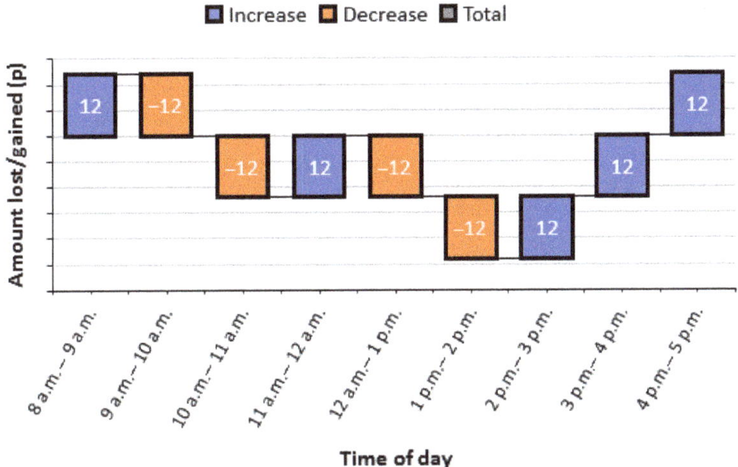

Figure 5. Campus net profit while exclusively pays off-peak prices.

Table 2 presents a 6.66 kW/h charger that brings in a net profit of 79.92 p/day when the campus pays off-peak prices at all times of the day.

The battery capacity needed to match 10 charging stations costs £27,432.90 depending on the supplier, whereas the 10 charging stations purchase and installation costs £8790, giving the same amount of energy storage at one time, assuming maximum capacity of EVs. This provides a saving of £18,642.90. Lithium-ion batteries have a lifespan of 2–3 years or 300–500 charge cycles meaning this cost will build up over time. The charging stations are designed to last at least 10 years with parts that are easily replaceable, and most of them have a warranty of 3 years. The charging and discharging rate of a 6.66 kW/h charger throughout the day has been presented in Figure 6. The outcome of charging and discharging the EV battery by 6.66 kW/h using a 6.66 kW charger if the car leaves with the same capacity as it was once plugged in, the EV owner could earn 79.92 p/day. If the EVs' owners return to their vehicle and it is out of charge, they are unlikely to use the V2G method. Taking this into account, for this simulation, the campus must have the car fully charged by 17:00 when most classes are finished.

Table 2. Net profit per day while the campus pays off-peak prices.

Time	Price (Pence)	Charge/Discharge	EV Owner (Pence)	Campus (Pence)	Campus at Off-Peak Prices (Pence)
08:00–09:00	79.92	Charge	−79.92	+79.92	+79.92
09:00–10:00	99.99	Discharge	+99.9	−99.9	−79.92
10:00–11:00	99.99	Discharge	+99.9	−99.9	−79.92
11:00–12:00	79.92	Charge	−79.92	+79.92	+79.92
12:00–13:00	99.99	Discharge	+99.9	−99.9	−79.92
13:00–14:00	99.99	Discharge	+99.9	−99.9	−79.92
14:00–15:00	79.92	Charge	−79.92	+79.92	+79.92
15:00–16:00	79.92	Charge	−79.92	+79.92	+79.92
16:00–17:00	79.92	Charge	−79.92	+79.92	+79.92
		Net profit	0	0	79.92

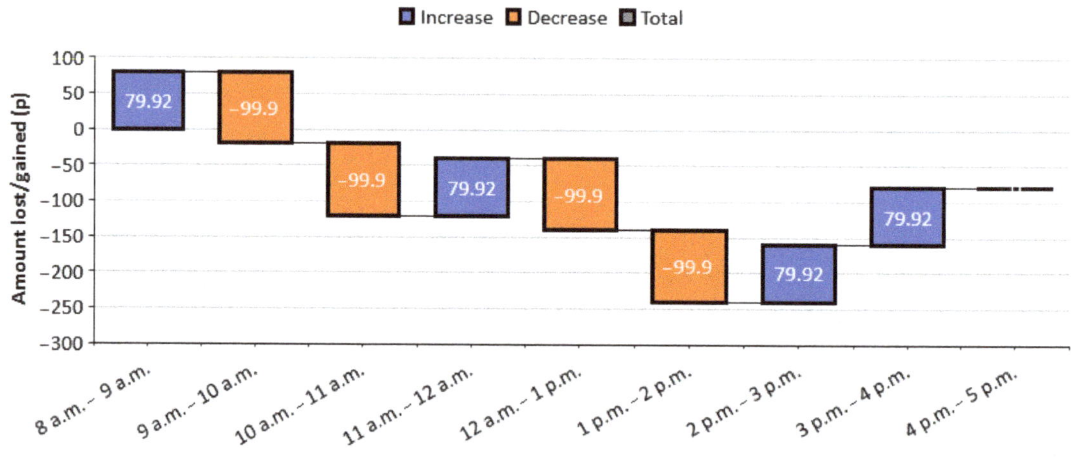

Figure 6. The profit earned by the campus per day.

The EV owner could earn 79.92 p per day, but the campus would lose that too. The campus, however, would not have to pay for large battery installation for the energy storage. In the case of 10 charging stations, the approximate loss becomes £7.99 a day for the university. The loss may accumulate up to the total cost of the battery capacity of 10 charging stations after 6.4 years. The battery must be changed every 2–3 years depending on the frequency of use. Overall, if the university campus installed ten 6.66 kW/h EV charging stations instead of buying the lithium-ion battery capacity to cover the capacity of the charging stations, the university would save £79,856.29 every 10 years. It is assumed that the batteries had to be replaced every two years and the charging stations are every 10 years. The day was simulated and mapped out, assuming the car was plugged in at 08:00 at 80% and unplugged at 17:00 at 100%, as is shown in Figure 7.

This method charges the battery as much as possible, which is 6.66 kW/h for 5 h in the day and thus, discharges it, so it sums to 100% by 17:00. This was calculated through:

$$Dc = Oc + Cr - Fc = \ 32\ kW + 33.3\ kW\ \ - 40\ kW\ \ = 25.3\ kW \quad (11)$$

In Equation (11), the 'Dc' is the discharge/day, the 'Oc' is the charge before the EV is plugged in, the 'Cr' is how much the EV is charged during a cycle, and the 'Fc' is the EV's capacity at 100% charge. This allows the EV to leave with 100% charge, while still discharging as much as possible at peak times.

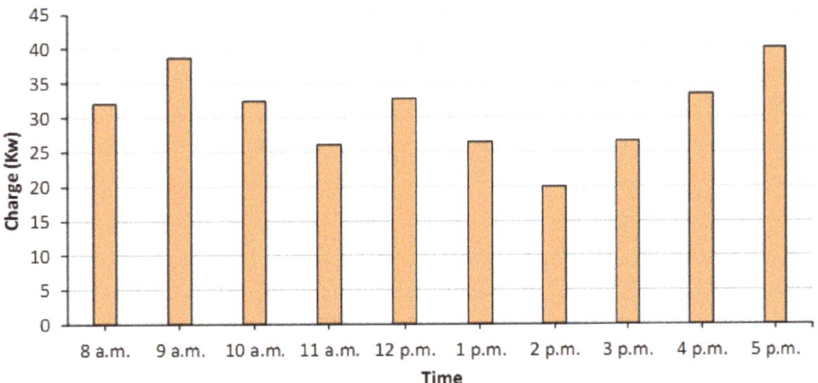

Figure 7. The times of peak and off-peak while the battery is being charged or discharged.

3.3. Net Savings and Charge and Discharge Times

The savings for the owner is largely dependent on the cost of the battery for the EV as a costly battery may lead to a lower net profit. The savings can be calculated through this section. A new Tesla Model S has a battery warranty of 8 years, which can last 20 years or more [12]. Three EV models are used to calculate the rate of battery degradation. The models are classified as type A, B and C. EV type A, and type B have a battery warranty of 8 years. The cost of type C's battery is replaced roughly $5000 to $7000 according to 'Interesting Engineering' [11], whereas type A's battery will cost $5500 [13] with a lifespan of 20 years also. On average, the cost for an EV car battery then is $6000 or £4666.20 on 20 February 2020. The lifespan of an EV battery during this method must include faster battery degradation, so using the data taken from References [11,12,36,37], it would have a rough lifespan of 4.81 years until it has 80% of its full battery capacity left, which has been shown below.

If the average UK mileage is 12,231 km per annum and type A's battery lasts 20 years, this equates to 244,620 km. The term '*NFc*' is the number of full charges, '*DL*' is the distance travelled before 20% battery capacity loss, and '*DFc*' is the distance from a full charge in miles.

$$NFc = \frac{DL}{DFc} = \frac{152,000}{73} = 2082 \quad (12)$$

Assuming the EV is plugged in at 80% at 08:00 and is un-plugged at 100% at 17:00, the battery discharges by 25.3 kW/day, and charges 33.3 kW/day. The battery is discharged 63.25% per day. The standard capacity '*SC*' over 20 years is:

$$SC = NFc \times Fc = 2082 \times 40\ kW = 83,280\ kW \quad (13)$$

The EV must charge by 69,330.6 kW over the 20 years instead of 16,656 kW if it were not using the V2G method. If 16,626 kW reduces the battery to 80% in 20 years, then at 208.2 kW, there is a 1% loss per year. If the EV is being charged by 69,330.6 kW, there is a 4.16% loss per year. The lifespan calculation of the battery has been shown in Equation (14).

$$\frac{20\ Y}{4.16\%} = 4.81\ Y \quad (14)$$

The EV's battery will have deteriorated to 80% after 4.81 years of the maximum capacity where Tesla recommends that the battery is replaced. If EV type A's battery costs £4265 and needs to be replaced every 4.81 years instead of every 20, the incentives must be great. As shown above, the EV owners will make £268.52 /year or £1291.64 before purchasing another battery. For the consumer to break even, the campus must pay £3.30

per day or £0.825 per hour to the EV owner, assuming the simulation parameters. The campus will save £1749.45 per year using these parameters.

The outcome will be greatly affected by the energy demand of the building as for this example, if the battery storage were for only peak times, it would only need to have a capacity of 1916 kW over 4 h, which is equivalent to 72 charging stations. A 10-year simulation on the cost analysis and ROI of installing EV chargers on campus with storage equivalent of all times of the day showing V2G storage is significantly less costly in Figure 8.

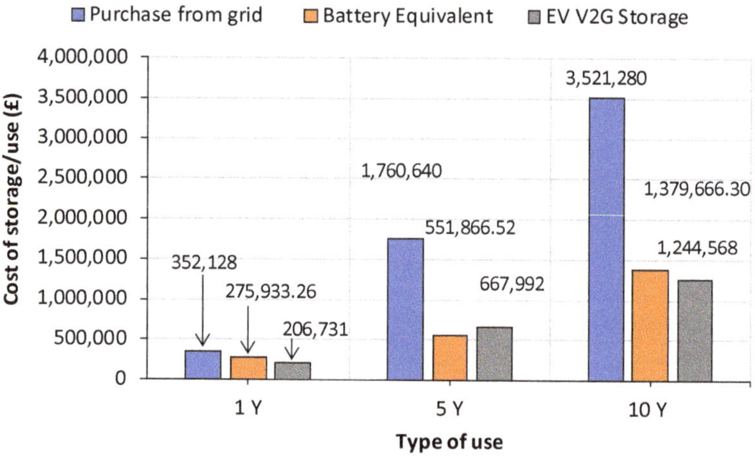

Figure 8. 10-year energy price for all times of the day.

The battery equivalent is only the price for the battery capacity, as the battery will also need to be charged, so the energy still needs to be bought. The EV V2G storage is the cost of the charging stations. After 2 years, the campus Li-ion battery should be replaced, whereas the EV battery should be replaced every 4.81 years. A 10-year simulation on the cost analysis and ROI of installing EV chargers on campus with storage equivalent of only peak-times of the day leading to a smaller cost for V2G storage than any other system, which is presented in Figure 9.

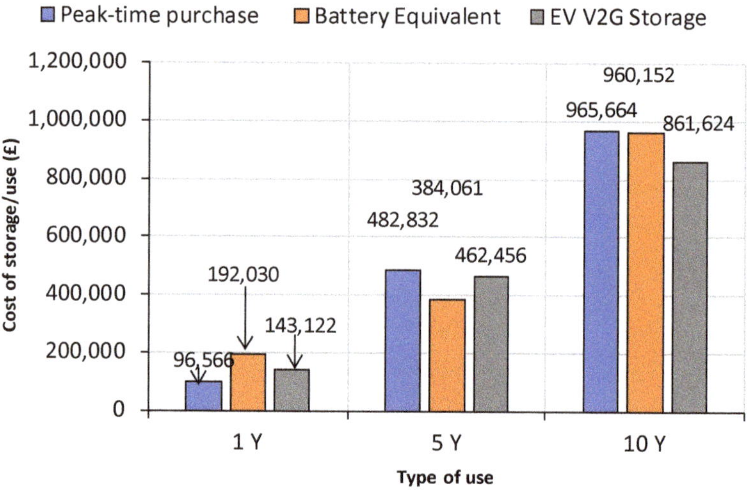

Figure 9. A 10-year simulation on the cost analysis during peak time.

A 10-year simulation of cost analysis and ROI of installing EV chargers on campus with storage equivalent of peak and off-peak times shows the significant price difference between off-peak and peak-time, which is presented in Figure 10.

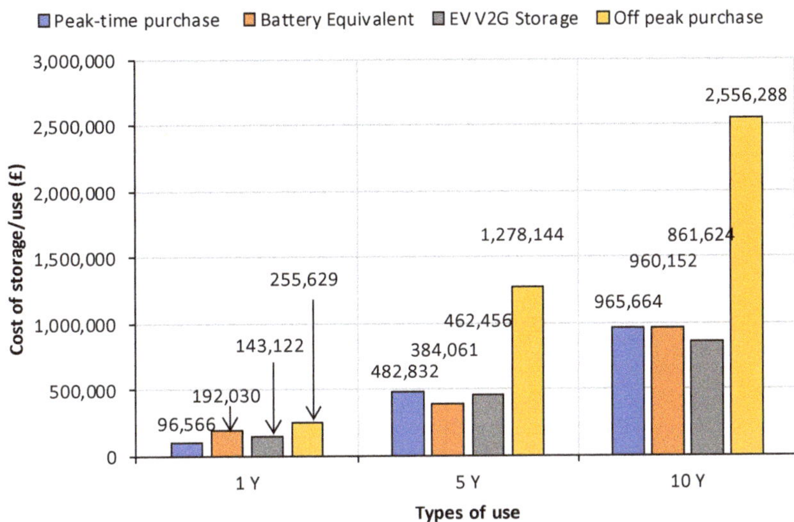

Figure 10. A 10-year simulation on the cost analysis on campus during off-peak time.

After one year, it is cheaper to purchase the peak-time energy from the grid, rather than using the V2G method at peak-times. This is because the EV chargers must be bought and installed for the first year. The year 5 and 10 year plans show that the V2G method is cheaper than purchasing from the grid. Off-peak prices from the grid are higher overall than peak-time as more off-peak hours are available than peak-hours.

3.4. Energy Consumption Prediction Using ML

While most research and applications of machine learning and computational intelligence techniques relate to the energy consumption and price of electricity, the use of such technologies for enhancing future prediction is yet to be realised and demonstrated. A neural network (NN) has been employed to predict the future energy demand and the future V2G cost for the years 2018 and 2019. The predictions for energy demand on each month in a year have been presented in Figure 11.

A general smooth trend is observed from January to April in Figure 11. However, the prediction accuracy has been slightly decreased in the month May for the year 2018 and 2019, due to the inconsistency of the trend in actual data observed in training compared to the previous years. The fundamental principles of ML techniques are to follow the trend of its training state during prediction. Hence, the prediction of energy consumption for the remaining months gradually increases in this study. The data showing as blue in the graph is indicated as actual data, whereas the orange data is the predicted output.

The yearly average prediction of energy demand for the building has also been presented in Figure 12 for the same years 2018 and 2019. It is observed that a smooth trend exists in the actual data for all years. Hence, the predictions followed the actual curve.

The predicted yearly average and actual recorded yearly average are different. The prediction error for 2018 and 2019 are 13,170 kW and 8846 kW. The error percentages for the years are 5.62% and 3.98%, respectively. This error is due to the volatility of energy consumption.

The data collected through electricity meters and the predicted data using the machine learning are varied. This variation can be shown as the prediction error below in Figure 13.

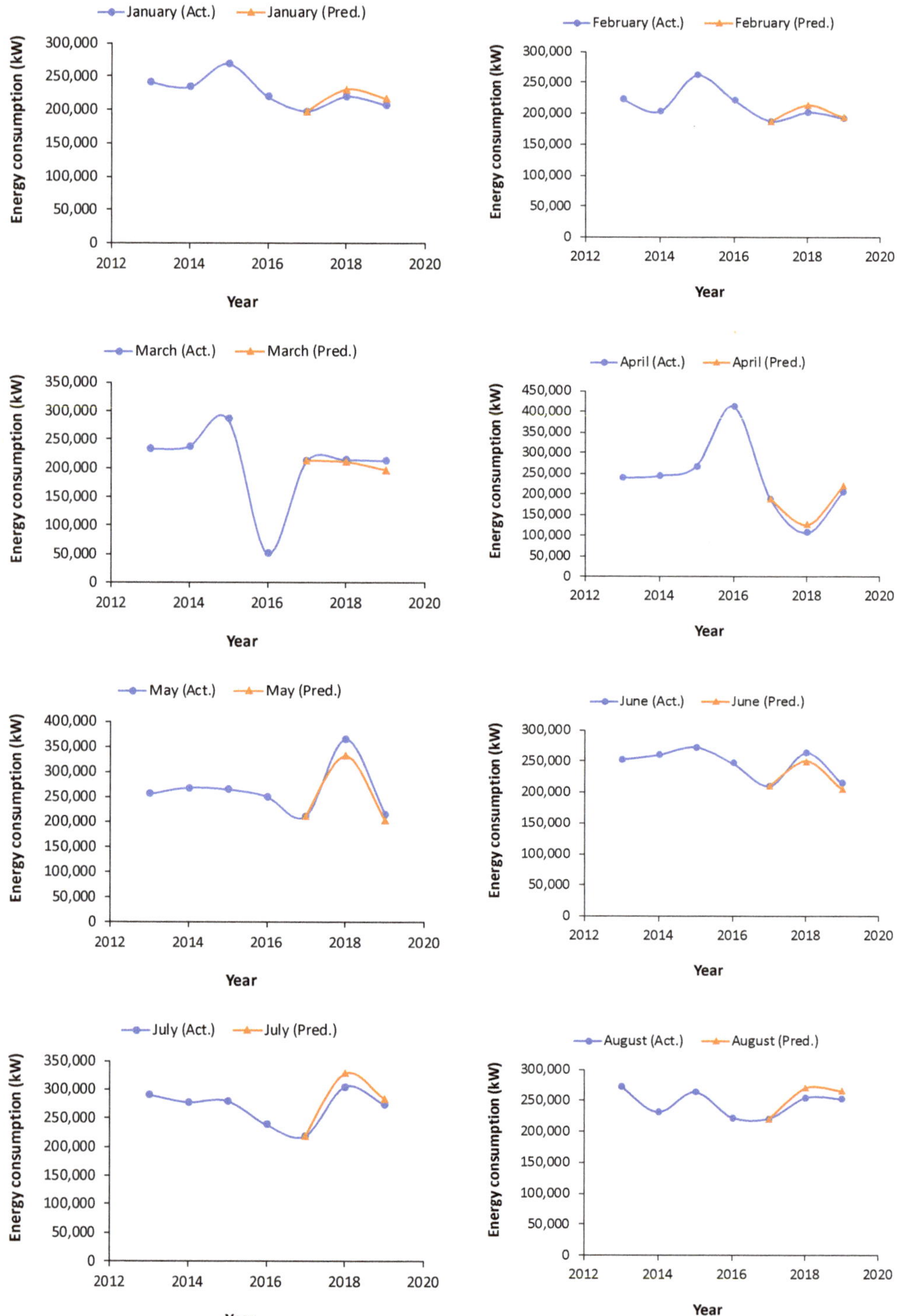

Figure 11. *Cont.*

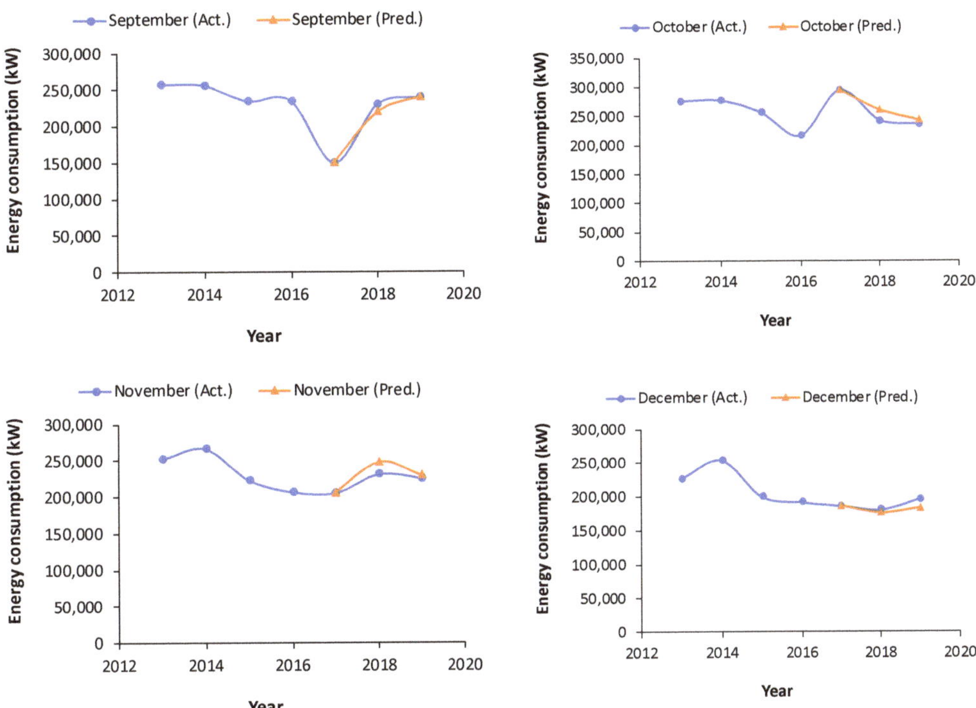

Figure 11. Energy consumptions prediction on a monthly basis for the year 2018 and 2019 using the MLA. (e.g., November (Act.) and November (Pred.).

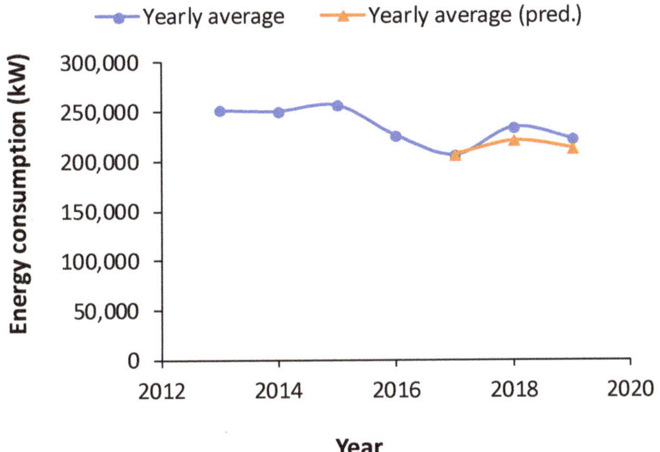

Figure 12. Annual average energy demand prediction using NN.

The largest error was in April 2018, at 15.1%. The month with the lowest error was in September 2019 at 0.08%. April's recorded energy consumption variance was the highest, at 98,162 kW. September's recorded energy consumption variance was among the lowest, at 9680 kW. The input data consisted of the energy consumption, so the more volatile the data, the harder it is to predict, and thus, the inaccuracy for the month, April. The average prediction error across all outputs is 5.1%.

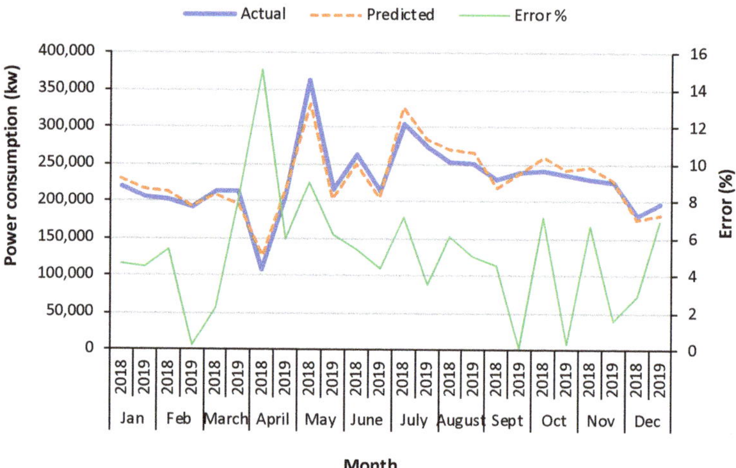

Figure 13. Error percentage for predicted energy consumption compared to recorded.

3.5. *Cost of Electricity Prediction for V2G Using ML*

The employment of NN has brought great performance in predicting the cost of electricity, which has been presented in Figure 14. From the results, it is found a smooth prediction trend for the month January and February. However, the prediction accuracy has been observed irregularity for the month May to July, due to having nonsmooth trend of the actual data for those months for the year 2018 and 2019, which has been shown as yellow colour in the figure. The prediction cost from the month August to December has shown a smooth trend between the actual and predicted values. Although accuracy has been slightly decreased for the months September and October.

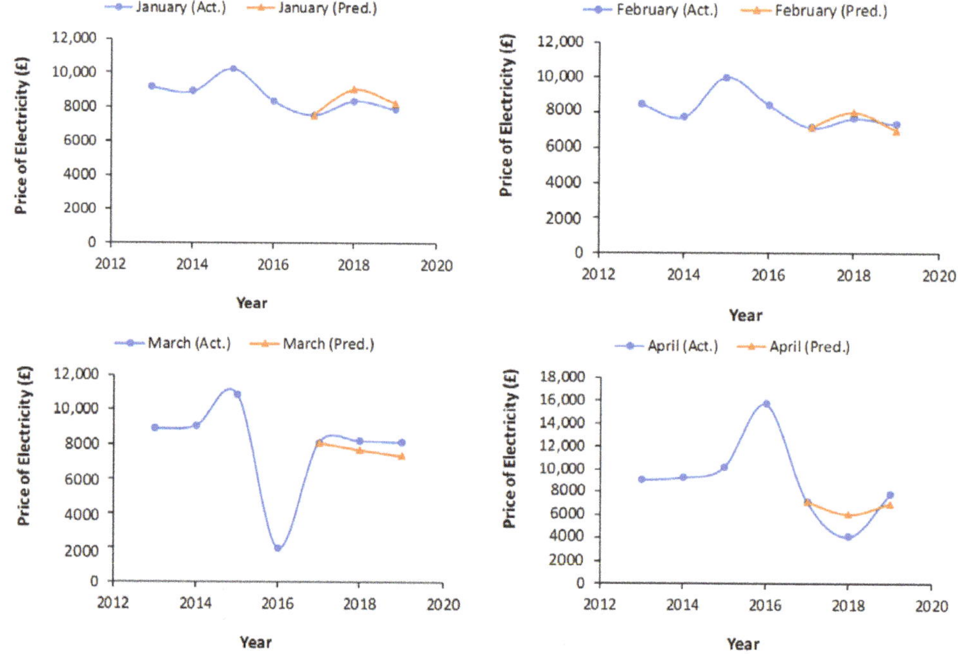

Figure 14. *Cont.*

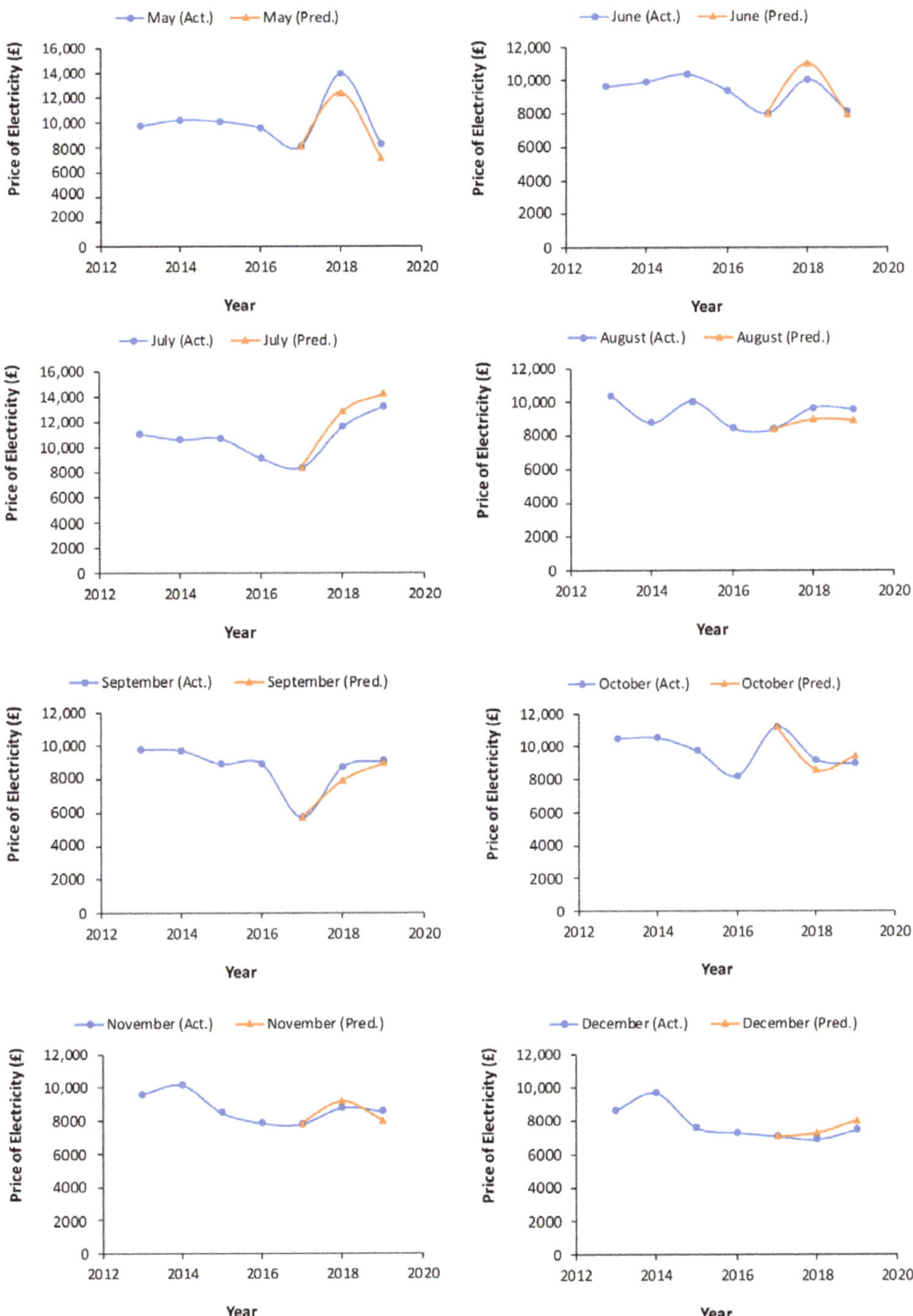

Figure 14. The prediction of cost for electricity of V2G on a monthly basis for the year 2018 and 2019 using ML.

The yearly average cost of the building has also been presented in Figure 15 for the year 2018 and 2019. Having the smooth trend of actual data for all years, the prediction for both the years followed the actual trend.

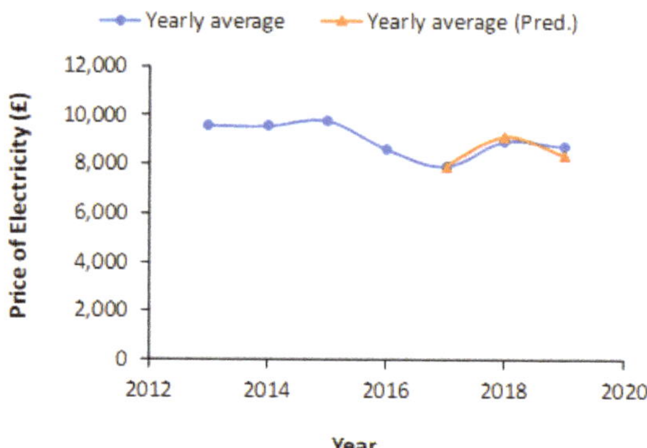

Figure 15. Annual average cost of electricity prediction using NN.

The months in the middle of Figure 16, within summer, have a higher power consumption than months in neighbouring seasons. While building becomes hotter, air conditioning is used instead of heating. This shows that air conditioning uses more energy than heating for the building. The power consumption varies from roughly 225,000 kW to 290,000 kW.

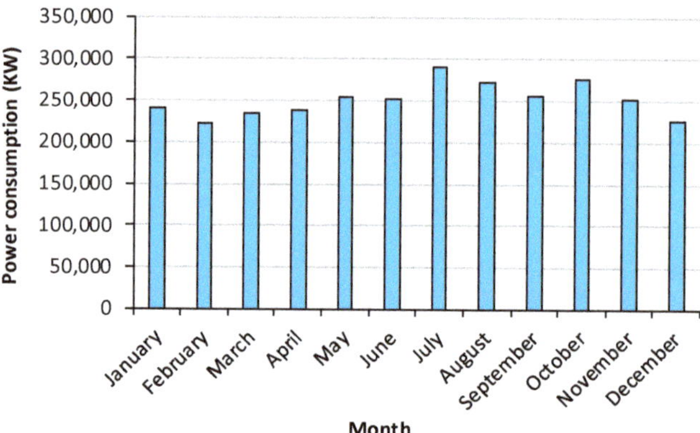

Figure 16. Monthly basis energy consumption of the building.

The data taken from electricity meters and the predicted data through ML are different. In the case of larger difference, the error is shown high (Figure 17).

The largest error percentage was 32% in April 2018. The month with the lowest error was in September 2019 at 1.74%. The average error in prediction is 7.94% over 2018 and 2019. April is more varied through the years than the other months. It ranges from £15,662 to £4081 for the use of the V2G method between 2016 and 2019. This is a difference of £11,581. September is less varied. It ranges from £5711 to £9105. This is a difference of £3393. The average error across all months is 7.9%. The more varied the data is, the more data is necessary for an accurate prediction. Between 2017 and 2019, April's calculated V2G cost varied by £6822, whereas between 2017 and 2019, it varied by £5823. The cost of the V2G method is directly linked to energy consumption. The variance of the date is the reason for the error. Additional data, including weather, footfall, etc., can be added to enhance the performance of ML.

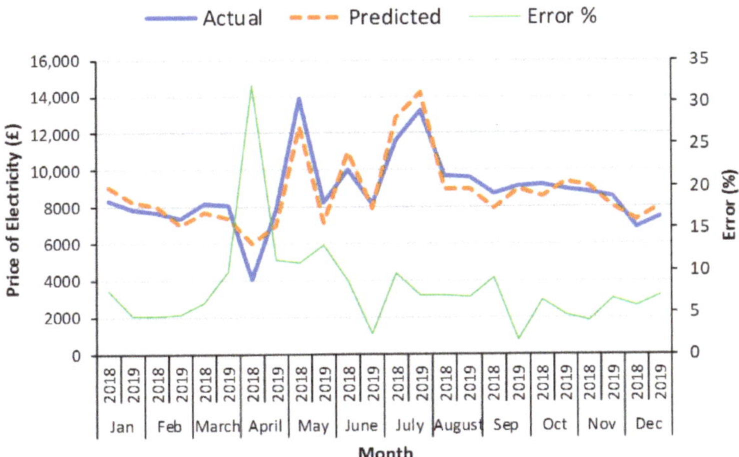

Figure 17. Error percentage for predicted EV purchasing price.

3.6. General Discussion

To cover the cost of battery degradation for the EV owner, the energy must be bought at 85.2 p/kW. This price has been considered to build ML models. Figure 11 has been analysed by comparing the actual data and the predicted data. The difference between the predicted and actual energy demand yearly averages, as shown in Figure 12, is a total of 22,016 kW between 2018 and 2019. This gives a prediction error of 2.07%. The least accurate month is in August, shown in Figure 11, the variance in 2018 and 2019 is 29,835 kW, which is 5.57% of the maximum value. The least accurate month from Figure 14, showing the price of the V2G method, is in April. On the other hand, the values collected through installed electric meters from 2018 and 2019 are £4081 and £7818, respectively, whereas the predicted values are £6004 and £6944, respectively. The prediction errors for 2018 and 2019 are £1923 and £875, which are 32% and 11%, respectively. This variance stems from the volatility of the energy demand in April, as is shown in Figure 11.

The prediction error is most common in months with a larger variance of data. The more varied and inconsistent the input data is the more input data is needed to secure an accurate output. In Figures 13 and 17, has the highest energy consumption, also the month with the highest price for V2G use, whereas the same with the months with the lowest V2G price and energy consumption. The method is used so that the EVs are fully charged by 17:00, allowing the EV owner to drive home and back to the university building the next day with an 80% charge, but this can be changed in the future. Smart metering can be used so the owner of the EV can input what time they will leave. The method will then be altered, depending on the leaving time, so the EV is fully charged. It is refreshed hourly; however, the time could be shortened to provide a more efficient V2G method.

In this work, a novel smart multienergy system with the ability to combine various energy storage technologies have been proposed to provide the best economic and environmental options for a given demand. The EV, energy storage and transaction of on-site energy must work in unison to enable an effective V2G model. Mazzoni et al. [38] have analysed the use of energy storage systems, including combined heat and power units, which show great financial and environmental benefits. V2G provides an incentive-pricing plan by motivating electric vehicles owners through participating in a charging/discharging system [39]. The master planning issue on this could be that the EVs' owners can charge vehicles at a low cost using unused or extra power in the grid during off-peak demand. In the case of shortage of power in the grid system during on-peak demand, EVs owners can earn money by discharging extra stored power from their vehicles at a higher price.

The implementation of a V2G method in any university campus or similar set-up requires inspection of key economic parameters, including initial investment, operational expenditure, maintenance, return on investment, and end net profit. Key technical parameters of energy storage include type (thermal, chemical, kinetic), capacity, physical size, charge and discharge rate, depth of discharge, and lifespan of the storage technique [39]. The economic parameters dictate the transferability of the V2G method as EV chargers must be installed, and there must be a return on investment to confirm that the method is transferrable to another circumstance. The technical parameters dictate how effective the method is. The buildings' characteristics (useable space, times of use, demand etc.) need to be met with a battery of the correct size, capacity, depth of discharge, and rate of charge, to ensure the method is effective. The replicability of the V2G method is dependent on available space for EV chargers, energy characteristics of the building, initial investment and storage techniques.

4. Conclusions

A V2G set-up of a University campus was modelled based upon energy demands, potential supply and net profits. A method of integrating V2G technology into a campus has been created under different scenarios, various demands, such as peak time and off-peak, to allow the V2G method to take place. The investigation shows that the proposed method is economically and environmentally beneficial. The proposed method can adapt depending on the variables, such as the type of EV, battery life, times and intensity of supply and demand etc. Results reveal that using the charge from EVs costs 64.7% less and 9.79% less than purchasing from the grid and using battery storage, over the 10-year period, at all times of the day. For only peak times, the cost was 10.8% less and 10.3% less than purchasing from the grid and using battery storage over the 10-year period. The finance authority can change to match requirements to suit the owners of the installations or to add incentives for the EV users. Thus, both parties gain profit instead of simulating just the benefit of using V2G for the National Grid, which has been performed thoroughly and presented in various articles [9]. The university campus is used instead, showing the great financial benefits to provide incentives to improve the growth of the V2G utilisation.

The future of EVs is rapidly growing with efficiency, effectiveness [40–43], and popularity for reducing carbon footprint [44–47], lower cost of EVs, the opportunity of battery replacement and the increment of the lifespan of the batteries. Although many research works are conducting on surrounding EVs [48–50] however, the main obstruction for any V2G method is that the EV consumers will have to buy new batteries more often, and they will require an incentive to sell their electricity. Therefore, the electricity from the EVs needs to be paid for at a high enough rate that at least the replacement of the battery can be covered. This drastically reduces the effectiveness.

Feed forward NN has provided accurate results regarding the energy generation and V2G cost predictions, for the months of April and May. The months of April and May were less accurate, due to the volatility of the energy demand, with an error of 11% and 32%. More data regarding the V2G price could be added into the MLA, and various MLA's can be employed to prove the accuracy of volatile energy demands. The V2G price analysis didn't include the price of installation, only the price of purchasing the energy from the EVs at 85.2 p/kW. The main method used 104 EV chargers, whereas the MLA for the V2G price uses; however, many chargers are needed for the demand of the month. The amount of 6.66 kW chargers ranges from 24 to 104. Further work on the MLA could include integrating this with other energy factors of the building, such as renewable generation, to drive future building development towards carbon reduction.

In the future, a smart meter can be designed to meet the EV owners' demands so that it can be decided to select the time of charging their EVs. The users should be well informed about the degradation of their car battery and estimated their net profit. In addition, buildings with many customers, such as the Manchester University campus or various businesses, could reduce the cost of energy bill by applying the proposed method.

This method then relies on the research and improvement of EVs, specific to EV batteries and price, allowing the consumer to make a profit, giving the National Grid or any supplier more freedom for the variables of the method. Based on achieved results in this paper, the method entertains various techniques demonstrating a viable option for EVs in future.

Author Contributions: Conceptualization, C.S. and M.A.; methodology, C.S. and M.A.; validation, M.A. and A.A.; formal analysis, M.A. and A.A.; investigation, C.S., M.A. and A.A.; resources, A.A.; data curation, C.S., writing—original draft preparation, C.S., writing—review and editing, M.A. and A.A.; visualization, C.S., M.A. and A.A.; supervision, M.A. and A.A.; project administration, M.A. and A.A. All authors have read and agreed to the published version of the manuscript.

Funding: This research received no external funding.

Institutional Review Board Statement: Not applicable.

Informed Consent Statement: Not applicable.

Data Availability Statement: Not applicable.

Conflicts of Interest: The authors declare no conflict of interest.

References

1. Committee on Climate Change. How the UK Is Progressing. 2019. Available online: https://www.theccc.org.uk/ (accessed on 10 September 2020).
2. Seck, G.S.; Krakowski, V.; Assoumou, E.; Nadia Ma zi Marauric, V. Embedding power systems reliability within a long-term energy system optimization model: Linking high renewable energy integration and future grid stability for France by 2050. *Appl. Energy* **2020**, *257*, 114037. [CrossRef]
3. Squire Energy. Peak Power Plants Explained. Available online: https://squireenergy.co.uk/peak-power-plants-explained/ (accessed on 11 September 2020).
4. Society of Motor Manufacturers and Traders. Electric Vehicle and Alternately Fuelled Vehicle Registrations. 2020. Available online: https://www.smmt.co.uk/vehicle-data/evs-and-afvs-registrations/ (accessed on 15 September 2020).
5. Hirst, D. *Electric Vehicles and Infrastructure*; Number CBP07480; House of Commons Library: London, UK, 2020.
6. Elexon Portal, Sheffield University. G.B. National Grid Status. Available online: https://www.gridwatch.templar.co.uk/ (accessed on 25 September 2020).
7. Battery University. BU-1003: Electric Vehicle (EV). Available online: https://batteryuniversity.com/learn/article/electric_vehicle_ev (accessed on 20 September 2020).
8. Spirit Energy. Domestic EV Charging. 2020. Available online: https://www.spiritenergy.co.uk/ev-charging (accessed on 27 September 2020).
9. Li, X.; Tan, Y.; Liu, X.; Liao, Q.; Sun, B.; Cao, G.; Li, C.; Yang, X.; Wang, Z. A cost benefit analysis of V2G electric vehicles supporting peak shaving in Shanghai. *Electr. Power Syst. Res.* **2020**, *179*, 106058. [CrossRef]
10. Hull, R. How Long Will Electric Car Batteries Really Last? Study Claims Teslas Lose Just 1% Performance Every Year Caused by Repeat Charges. 2019. Available online: https://www.thisismoney.co.uk/money/cars/article-7764529/US-study-claims-Tesla-batteries-lose-just-1-performance-year.html (accessed on 24 September 2020).
11. Yurday, E. Average Car Mileage UK. 2020. Available online: https://www.nimblefins.co.uk/average-car-mileage-uk (accessed on 21 October 2020).
12. Loveday, S. Nissan LEAF Batteries to Outlast Car by 10–12 Years. 2019. Available online: https://insideevs.com/news/351314/nissan-leaf-battery-longevity/ (accessed on 28 September 2020).
13. Çeven, S.; Albayrak, A.; Bayir, R. Real-time range estimation in electric vehicles using fuzzy logic classifier. *Comput. Electr. Eng.* **2020**, *83*, 106577. [CrossRef]
14. Lambert, F. A Look at Tesla Battery Degradation and Replacement After 400,000 Miles. 2020. Available online: https://electrek.co/2020/06/06/tesla-battery-degradation-replacement/ (accessed on 10 February 2021).
15. Solanke, T.U.; Ramachandaramurthy, V.K.; Yong, J.Y.; Pasupuleti, J.; Kasinathan, P.; Rajagopalan, A. A review of strategic charging-discharging control of grid-connected electric vehicles. *J. Energy Storage* **2020**, *28*, 101193. [CrossRef]
16. Shi, R.; Li, S.; Zhang, P.; Lee, K.Y. Integration of renewable energy sources and electric vehicles in V2G network with adjustable robust optimization. *Renew. Energy* **2020**, *153*, 1067–1080. [CrossRef]
17. Kester, J.; de Rubens, G.Z.; Sovacool, B.K.; Noel, L. Public perceptions of electric vehicles and vehicle-to-grid (V2G): Insights from a Nordic focus group study. *Transp. Res. Part D Transp. Environ.* **2019**, *74*, 277–293. [CrossRef]
18. Thompson, A.W.; Perez, Y. Vehicle-to-everything (V2X) energy services, value streams, and regulatory policy implications. *Energy Policy* **2020**, *137*, 111136. [CrossRef]

19. Li, J.; Yang, Q.; Co, W.; Li, S.; Lin, J.; Huo, D.; He, H. An improved vehicle to the grid method with longevity management in a microgrid application. *Energy* **2020**, *198*, 117374.
20. Qian, F.; Gao, W.; Yang, Y.; Yu, D. Economic optimization and potential analysis of fuel cell vehicle-to-grid (FCV2G) system with large-scale buildings. *Energy Convers. Manag.* **2020**, *205*, 112463. [CrossRef]
21. King, C.; Datta, B. EV charging tariffs that work for EV owners, utilities and society. *Electr. J.* **2018**, *31*, 24–27. [CrossRef]
22. Shea, R.P.; Worsham, M.O.; Chiasson, A.D.; Kissock, J.K.; McCall, B.J. A lifecycle cost analysis of transitioning to a fully electrified, renewably powered, and carbon-neutral campus at the University of Dayton. *Sustain. Energy Technol. Assess.* **2020**, *37*, 100576. [CrossRef]
23. Du, G.; Zou, Y.; Zhang, X.; Liu, T.; Wu, J.; He, D. Deep reinforcement learning based energy management for a hybrid electric vehicle. *Energy* **2020**, *201*, 117591. [CrossRef]
24. Nuvve. E-Flex Innovate UK. 2021. Available online: https://nuvve.com/projects/eflex-innovate-uk/ (accessed on 16 February 2021).
25. Christensen, B. Parker's Vehicle-Grid Integration Summit Was Sold Out. 2018. Available online: https://parker-project.com/parkers-vehicle-grid-integration-summit-was-sold-out-when-v2g-results-were-presented/ (accessed on 16 February 2021).
26. Making Vehicle-to-Grid a Reality: V2G Developments in the UK. Available online: https://parker-project.com/wp-content/uploads/2018/12/VGI-Summit-Day-2-S1-Initiatives-DK_2018-Vehicle-to-Grid_UK.pdf (accessed on 12 November 2020).
27. EVANNEX. Just How Long Will an EV Battery Last? 2019. Available online: https://insideevs.com/news/368591/electric-car-battery-lifespan/ (accessed on 29 September 2020).
28. PodPoint. Smart Home Charger. 2020. Available online: https://pod-point.com/products/homecharge (accessed on 30 September 2020).
29. Kotsiantis, S.B.; Zaharakis, I.D.; Pintelas, P.E. Machine learning: A review of classification and combining techniques. *Artif. Intell. Rev.* **2006**, *26*, 159–190. [CrossRef]
30. Vercellis, C. *Business Intelligence: Data Mining and Optimization for Decision Making*; Wiley: New York, NY, USA, 2009; pp. 1–420.
31. Gorunescu, F. *Data Mining: Concepts, Models and Techniques*; Springer: Berlin/Heidelberg, Germany, 2011; Volume 12.
32. Gao, Y.; Wang, J.; Chen, H.; Li, G.; Liu, J.; Xu, C.; Huang, R.; Huang, Y. Machine learning-based thermal response time ahead energy demand prediction for building heating systems. *Appl. Energy* **2018**, *221*, 16–27. [CrossRef]
33. Wang, Z.; Hong, T.; Piette, M.A. Building thermal load prediction through shallow machine learning and deep learning. *Appl. Energy* **2020**, *263*, 114683.
34. Liu, C.; Sun, B.; Zhang, C.; Li, F. A hybrid prediction model for residential electricity consumption using holt-winters and extreme learning machine. *Appl. Energy* **2020**, *275*, 115383. [CrossRef]
35. Engelbrecht, A.P. *Computational Intelligence—An Introduction*; John Wiley & Sons Ltd.: West Sussex, UK, 2002.
36. Ahsan, M.; Stoyanov, S.; Bailey, C.; Albarbar, A. Developing computational intelligence for smart qualification testing of electronic products. *IEEE Access* **2020**, *8*, 16922–16933. [CrossRef]
37. Soumeur, M.A.; Gasbaoui, B.; Abdelkhalek, O.; Ghouili, J.; Toumi, T.; Chakar, A. Comparative study of energy management strategies for hybrid proton exchange membrane fuel cell four-wheel drive electric vehicle. *J. Power Sources* **2020**, *462*, 228167. [CrossRef]
38. Mazzoni, S.; Ooi, S.; Nastasi, B.; Romagnoli, A. Energy storage technologies as techno-economic parameters for master-planning and optimal dispatch in smart multi energy systems. *Appl. Energy* **2019**, *254*, 113682. [CrossRef]
39. Bibak, B.; Tekiner-Moğulkoç, H. A comprehensive analysis of Vehicle to Grid (V2G) systems and scholarly literature on the application of such systems. *Renew. Energy Focus* **2021**, *36*, 1–20. [CrossRef]
40. Fachrizal, R.; Munkhammar, J. Improved photovoltaic self-consumption in residential buildings with distributed and centralized smart charging of electric vehicles. *Energies* **2020**, *13*, 1153. [CrossRef]
41. Taljegard, M.; Göransson, L.; Odenberger, M.; Johnsson, F. Electric vehicles as flexibility management strategy for the electricity system—A comparison between different regions of Europe. *Energies* **2019**, *12*, 2597. [CrossRef]
42. Zweistra, M.; Janssen, S.; Geerts, F. Large scale smart charging of electric vehicles in practice. *Energies* **2020**, *13*, 298. [CrossRef]
43. Alimujiang, A.; Jiang, P. Synergy and co-benefits of reducing CO2 and air pollutant emissions by promoting electric vehicles—A case of Shanghai. *Energy Sustain. Dev.* **2020**, *55*, 181–189. [CrossRef]
44. Hu, Y.; Wang, Z.; Li, X. Impact of policies on electric vehicle diffusion: An evolutionary game of small world network analysis. *J. Clean. Prod.* **2020**, *264*, 121703. [CrossRef]
45. Wang, M.; Tian, Y.; Liu, W.; Zhang, R.; Chen, L.; Luo, Y.; Li, X. A moving urban mine: The spent batteries of electric passenger vehicles. *J. Clean. Prod.* **2020**, *264*, 121769. [CrossRef]
46. Liu, X.; Sun, X.; Li, M.; Zhai, Y. The effects of demonstration projects on electric vehicle diffusion: An empirical study in China. *Energy Policy* **2020**, *139*, 111322. [CrossRef]
47. Dong, X.; Zhang, B.; Wang, B.; Wang, Z. Urban households' purchase intentions for pure electric vehicles under subsidy contexts in China: Do cost factors matter? *Transp. Res. Part A Policy Pract.* **2020**, *135*, 183–197. [CrossRef]
48. Nimesh, V.; Sharma, D.; Reddy, V.M.; Goswami, A.K. Implication viability assessment of shift to electric vehicles for present power generation scenario of India. *Energy* **2020**, *195*, 116976. [CrossRef]
49. Michaelides, E.E. Thermodynamics and energy use of electric vehicles. *Energy Convers. Manag.* **2020**, *203*, 112246. [CrossRef]
50. Khan, S.A.; Kadir, K.M.; Mahmood, K.S.; Alam, I.I.; Kamal, A.; Al Bashir, M. Technical investigation on V2G, S2V, and V2I for next generation smart city planning. *J. Electron. Sci. Technol.* **2020**, *17*, 100010. [CrossRef]

Article

A Circular Economy for the Data Centre Industry: Using Design Methods to Address the Challenge of Whole System Sustainability in a Unique Industrial Sector

Deborah Andrews [1,]*, Elizabeth J. Newton [2], Naeem Adibi [3], Julie Chenadec [4] and Katrin Bienge [5]

1 School of Engineering, London South Bank University, London SE1 0AA, UK
2 School of Applied Sciences, London South Bank University, London SE1 0AA, UK; liz.newton@lsbu.ac.uk
3 WeLOOP, 62750 Loos-En Gohelle, France; n.adibi@weloop.org
4 Digital Infrastructure Association (SDIA), 20354 Hamburg, Germany; julie.chenadec@sdialliance.org
5 Wuppertal Institute for Climate, Environment and Energy, 42103 Wuppertal, Germany; katrin.bienge@wupperinst.org
* Correspondence: deborah.andrews@lsbu.ac.uk

Abstract: The data centre industry (DCI) has grown from zero in the 1980s, to enabling 60% of the global population to be connected in 2021 via 7.2 million data centres. The DCI is based on a linear economy and there is an urgent need to transform to a Circular Economy to establish a secure supply chain and ensure an economically stable and uninterrupted service, which is particularly difficult in an industry that is comprised of ten insular subsectors. This paper describes the CEDaCI project which was established to address the challenge in this unique sector; this ground-breaking project employs a whole systems approach, Design Thinking and the Double Diamond methods, which rely on people/stakeholder engagement throughout. The paper reviews and assesses the impact of these methods and project to date, using quantitative and qualitative research, via an online sectoral survey and interviews with nine data centre and IT industry experts. The results show that the project is creating positive impact and initiating change across the sector and that the innovative output (designs, business models, and a digital tool) will ensure that sectoral transformation continues; the project methods and structure will also serve as an exemplar for other sectors.

Keywords: circular economy; design methods; whole systems thinking; data centre industry

Citation: Andrews, D.; Newton, E.J.; Adibi, N.; Chenadec, J.; Bienge, K. A Circular Economy for the Data Centre Industry: Using Design Methods to Address the Challenge of Whole System Sustainability in a Unique Industrial Sector. *Sustainability* **2021**, *13*, 6319. https://doi.org/10.3390/su13116319

Academic Editor: Luis Jesús Belmonte-Ureña

Received: 30 March 2021
Accepted: 26 May 2021
Published: 2 June 2021

1. Introduction

Data centres are interior spaces that house electronic and electrical equipment that processes and stores digital data. The data centre industry (DCI) has evolved from zero in the 1980s into a global industry with 7.2 million sites [1] in 2021, annual electricity consumption is 200 TWh (i.e., 1% of global electricity consumption) [2,3] and carbon emissions are equivalent to those from the pre-Covid airline industry [4]; 60% of the global population are now connected via smart phones, laptops, and other computing equipment [5], and the sector currently processes around 4.2 trillion gigabytes of data per year. It is predicted that the DCI will grow by around 500%, by 2030, as more people and objects are connected via the Internet of Things (IoT) [6]. Reliance on digital communications technology and, therefore, the DCI has developed to the point where any systems failure will adversely affect major commercial, health, education, and other sectors and dependence has been highlighted during the Covid pandemic when there was a notable increase in data traffic in response to remote working, education, and communication. The sector has helped to transform all aspects of life and has already proved very beneficial to many of those who have access to digital technology.

The speed and scale at which data centre technology, and equipment, developed far exceeded that of recycling infrastructure which, in conjunction with the fact that products were not designed with any consideration for treatment at end-of-life (i.e., linear design)

and poor perception of second life products, means that the DCI is contributing to the increasing volume of e-waste that is generated every year. Critical Raw Materials are essential to DC and other electronic equipment and many are lost to landfill, and/or incineration, or cannot be accounted for at end-of-life. Consequently, unless there is a change in practice across all equipment life cycle stages, there is a threat to the supply chain for these and other materials, which will destabilise the market, the DCI, and all services that rely on it. Furthermore, current practice is environmentally and socially unsustainable.

There is an urgent need to transform the sector, but this is limited by various technical and behavioural barriers. The CEDaCI project is the first of its kind, and it was set up to instigate change and develop a Circular Economy for the Data Centre Industry by using a whole systems approach in a fragmented industry, comprised of 10 sub-sectors, that operate in individual silos. A whole systems approach is essential because decisions made, and actions taken, at each life cycle stage affect the sustainability of all other life cycle stages. Design thinking and design methods encourage holistic thinking, and, therefore, they were selected as models for CEDaCI; they underpin the structure and process to ensure that the project delivers what the sector needs. People/stakeholders and a 'hearts and minds' ethos are central to both methods; considering the history of and work practice across the DCI, however, there was a risk of failure if representatives did not engage with the various events and platforms. The project launched in October 2018, and in February 2021, a survey of DCI representatives was carried out to ascertain the extent to which the project structure and methods were working, to learn whether amendments or revisions were necessary to ensure success, and to assess the impact of the project to date.

This paper first describes the context of and barriers to a Circular Economy in more depth. It then describes the design methods and documents, the project structure, activities, and content, followed by details and results of a quantitative online survey and qualitative data gathered via semi-structured interviews. Participant numbers were limited to 44 (with 32 usable data sets) and 9 respectively. The response rate appears to be associated with the COVID-19 pandemic, which significantly increased data traffic, workloads and absence due to ill-health. The results were very positive and showed that the project methods and process were appropriate, although there were some shortcomings due to lack of representatives from two subsectors; the methods and process are raising awareness of the challenges and potential solutions in the DCI, and of circularity in general, as well as facilitating development of digital tools, business models, CE-fit designs and prototypes, and new recycling and CRM reclamation processes. Finally, these results endorse the project as an exemplar and the methodology could be adapted and applied in other industrial sectors facing similar challenges.

2. Literature Review: Computing and Data Centres Past and Present

2.1. The Evolution and Expansion of Computing and Connectivity

The history of computing is believed to have started around 2000 years ago with the Antikythera Mechanism, which was discovered in the sea near Greece in 1901; it is described as a mechanical computer that was used to make astronomical predictions [7]. Computing and calculation machines and methods were also developed in the China, India, and the Islamic worlds prior to, and during, the mediaeval period. It was not until the 1820s, however, that modern computing began in the UK when Charles Babbage developed his mechanical calculation machines-the Difference Engine and the Analytical Engine—with support from Ada Lovelace; she was the first person to recognise that a universal computer could do anything providing that it was given the right data and instructions, and, therefore, she is often referred to as 'the first programmer' [8]. Development continued slowly through the 19th and early 20th centuries with the introduction of electro-mechanical machines and, in 1937, there was a significant conceptual breakthrough when the British mathematician Alan Turing published a paper describing an imaginary machine that performed simple mathematical tasks by following precise logical steps [9]. He continued to develop this concept and machines throughout World War II and, with colleagues at

GCHQ, achieved a significant technical breakthrough in 1942 when thermionic valves (generally used in telephone exchanges) were included in the machines. These components accelerated processing speed considerably and facilitated development of Colossus, the world's first completely programmable, electronic, digital computer, in 1943. Other notable parallel developments include the German Z series developed by Konrad Zuse (who used binary language to produce the first electronic calculator in 1938 and the first high-level programming language 1943–1945) and ENIAC (Electronic Numerical Integrator and Computer), which was developed and launched, in the USA, in 1945 for military use. It is worth noting that neither the German or British governments recognised the importance of, nor funded, Turing or Zuse's research at the beginning of WWII and, consequently, the antecedents to and early Colossus computers were made from recycled components discarded by telephone exchanges for example. Nevertheless, the various technical and programming developments continued at pace after WWII, and numerous computers were developed for industrial, commercial, and military applications. Since then, computing technology has developed ever more rapidly; for example, keyboard input capability and transistor-based technologies were introduced in the late 1950s. As the evolution of mainframe computing continued, other developments enabled the introduction of personal computing. Examples include the invention of integrated circuits, silicon-based transistors in the late 1950s, the first single-person operated computer (1962), the first microprocessor (1973), the first Apple 1 and 2 computers (1977), the first IBM personal computer (1982), the first Apple Macintosh (1984), and the first portable computers in the early 1990s. The first computer game was developed in 1961 at MIT, after which various dedicated affordable computers were developed and sold from 1979; computer games were important because they helped to introduce the wider public to and popularise computers, which ensured their place outside the workplace and in the home [10,11].

In 1965 Gordon Moore (an engineer and founder of Intel) predicted that, as density increased, the number of transistors that would fit on a computer chip would double every year—i.e., Moore's Law. Although this was amended to two years in 1975, this particular technical development, in conjunction with the development of new computer languages and software programmes, further accelerated the power and speed of computing; the development of data storage media, such as floppy discs and CDs, also increased functionality in the workplace and at home.

The earliest computers were stand-alone machines but gradually, research activity in the USA, France and UK facilitated the development of internal networks to which they were connected for data exchange. External networks were also developed and initially used in business and academia with the launch of a global Joint Academic Network (JANET) in 1984 to enable rapid information exchange and collaboration. During the 1980s, the British engineer and computer scientist Sir Tim Berners-Lee also developed a new digital information and communication language and network, which subsequently evolved to become the World Wide Web in 1989. Since then, the user group has expanded from 'geeks,' researchers, and academics to the general public, and in January 2021, over 4.66 billion people and 60% of the global population were 'connected' via the internet; of these individuals, 97% own a smart phone, 64% a laptop or desktop PC, and 34% a tablet [5].

2.2. Data Centres

While 'devices' (desk and laptop computers and mobile phones) serve as human-digital data interfaces, the hidden, but critical, enabler of connectivity is data centres (DCs). These facilities may be cupboard-sized or, (like the largest hyper-scale data centres in the world), equivalent in area to 93 football pitches; they all house digital data processing, networking and storage (ICT) equipment. Such is the popularity of the internet that, since its launch, the number of DCs around the world has grown considerably with estimated floor at 180 million m^2; 10 million m^2 of which is in Europe, with 70% concentrated in North West Europe (NWE) [12].

There are three main types of centres: Enterprise (private facilities dedicated to supporting single organisations), Colocation data centres (where space, equipment, and/or data storage, and processing capability is rented to customers) and Cloud data centres (which are entirely owned and managed by companies who rent their virtual infrastructure to clients who run and manage their own applications and data). Cloud service providers include well-known brands such as Google, Facebook, Microsoft, and Amazon who have other business arms, as well as less publicly-known companies (e.g., Rackspace, OVHCloud, and Serverspace).

In 2015, there were over 8 million data centres around the world; this total has decreased to 7.2 million [1], although floor space and data traffic are continuing to rise, and by 2022, there will be 4.2 trillion gigabytes of data processed per year. Development of the Internet of Things (IoT) will further increase data traffic and, consequently, it is estimated that data centre services will increase 300% in Europe by 2025 and 500% globally over 2018 figures [6].

Data centre equipment emits heat and, in addition to electricity for data processing and storage, DCs consume electricity for cooling and other operations. Consequently in 2018 the sector accounted for approximately 1% (205 TWh) of global electricity use [2,3] and emitted as much CO_2 as the commercial airline industry [4].

Since inception, the main concern and focus of the DC industry has been 100% uninterrupted service for customers. The most significant changes are increasing operational energy efficiency, some of which derives from changes in hardware, component design, and software; other reductions in electricity consumption have been achieved by changing from electro-mechanical air cooling to liquid cooling. This form of cooling can be system-based (e.g., open water circuits with adiabatic coolers or cooling towers) and/or localised (on-chip or submissive/immersion cooling). Examples of the latter use oil-based dielectric fluids whereas the former often use water; water is also used for humidification in air handling units. Water use is not a panacea and, although it can reduce operational energy inputs, it also has an adverse impact on the environment because the water discharged from data centres includes contaminants such as biocide chemicals (used to control Legionella in pipework and heat exchangers), dissolved salts (like limescale in kettles) and acidic or alkaline PH correctors, copper and steel corrosion protectors, and metal particles [13,14] Water cooling also uses 0.2–0.8 L per kWh used [15] so the sector consumes the equivalent of 120,000 Olympic sized swimming pools per year, all of which contributes to global water stress. Another alternative to electro-mechanical air cooling is described as free cooling, which draws exterior air into the centre; in this case, cold air is beneficial and, consequently, a growing number of data centres, and hyperscale centres in particular, are located either near to or in the Arctic Circle. The combined use of these differing technologies means that the DCI has a significant impact on energy and resource consumption

In summary, computing and associated industries evolved very slowly until the late 1930s when there were some major conceptual developments. Development accelerated a little until the 1960s when it began to accelerate more rapidly in response to major technical innovations; development further accelerated in the 1970s and 1980s, and computers gradually became accepted in the domestic as well as academic and commercial environments. The World Wide Web was launched in 1989, and since then, the sector has expanded in scope, scale, and at speeds unlike any other sector in history; within 30 years, life for those with access to the computers and the internet (now 60% of the global population) has been transformed. Although many adverse impacts are associated with connectivity (e.g., cyber bullying and fake news) they are outweighed by the many diverse positive benefits. Examples include access to education in remote locations, micro-loans, and new business development among disadvantaged communities in Africa, video calls with friends and family, and on-line shopping during the COVID-19 pandemic. Industry is also benefitting from digital technologies, such as robotics, additive, and other new manufacturing techniques and quality monitoring all of which will grow concurrently with the IoT while information exchange has transformed logistics, transport, public, and private travel.

2.3. The Challenge of e-Waste

The speed and scope of development, change, and the wish to provide consumers with smaller, lighter, more portable products with access to more data and storage capability, increasing processing speeds, product convergence, and functionality have all significantly influenced the design and manufacture of physical products. In the majority of cases, they have been and are designed for the present and life in use, which is generally limited by the design and life span of operating systems and firmware rather than failure of physical components.

The wish to both fulfil, and drive, consumer expectations has also driven developments in data centre equipment design. As previously mentioned, DCs may be repurposed spaces in commercial premises, but an increasing number are purpose built; size depends on location but generally, DCs have few windows. These centres are full of racks with servers, (which are, fundamentally, computers without screens), routers, switching and telecommunications equipment, mains and backup power supplies (such as batteries and generators, which are used to ensure uninterrupted service in the event of mains power failure), and cooling systems. The equipment is mainly comprised of metal and plastic cases and housing, and electrical and electronic components, the majority of which are made from a mixture of materials. Since the invention of integrated circuits, transistors and silicon chips, components have also decreased in size, but the underlying design principles of the equipment have not really changed.

This practice, in conjunction with rapid evolution of the sector, means that, in addition to other electrical and electronic products and sectors, the DC industry is contributing to the growing volume of e-waste/WEEE (waste electrical and electronic equipment) produced every year; in 2020 this was around 54 million tonnes (equivalent to 7.3 kg per person) and, unless there is a major change in culture, behaviour and treatment of equipment at end of life, this will increase to 120 million tonnes per year by 2050 [16]. This growth has been, and is being exacerbated, by the design and manufacture of products and components themselves, which makes separation of sub-components and materials either very difficult or impossible; although there is some evidence of steel, aluminium, and copper (i.e., larger components) recycling and gold recycling (which is found in relatively high concentrations in mobile phones and computers), the relatively low cost and size of the majority of individual components make recycling uneconomic unless executed at scale. It is impossible to say how much e-waste is actually recycled: while 17.4% is documented as collected and properly recycled, the remaining 82.6% cannot be accounted for; it is estimated that 8% is discarded in waste bins in high-income countries and that 7–20% is exported as second-hand products or e-waste to low-to-middle-income countries (LMICs) [16]. Although the exact whereabouts of the majority of e-waste is unknown, a considerable percentage is exported for processing and/or landfilling e.g., in Africa and China and similar countries, where practice and processes are frequently unregulated, often hazardous, and damage the environment and health.

Electrical and electronic equipment is complex and embodies approximately 69 elements: they include iron, aluminium and copper as well as precious metals (gold, silver, platinum, palladium, ruthenium, rhodium, iridium, and osmium), iron, aluminium, and copper as well as at least seven Critical Raw Materials (CRM). CRM are defined as such because they are economically and strategically important while availability is determined by concentration of production, geo-political location, potential of substitution, and current recycling rates. In 2011, the EU identified 14 such materials, but this increased to 20 in 2014, to 27 in 2017, and to 30 in 2020. They are already essential to renewable energy generation and low carbon technologies (such as wind turbines and batteries) and are becoming increasingly important to high technology products and emerging innovations [17,18], and like more mundane electrical and electronic products, demand is increasing.

Unlike older and other technologies, products, and materials (e.g., furniture, vehicles, clothing, mechanical equipment, electrical equipment), which can be repaired and recycled at end of life with comparative ease. As stated above, electronics do not; furthermore, recycled and resource reclamation infrastructure remains limited; there is also evidence of ring-fencing reserves, and China is purchasing mines and land in Africa, for example [19] as well as limiting export of rare earth minerals [20]. The combined impact of these factors is a threat to and potential disruptor of the supply chain, which will affect manufacturing capability unless manufacturers have materials reserves.

2.4. Data Centre Equipment Design and Manufacture

Data centre equipment life varies according to type, and while some last for 20 years, the average life of other products such as servers ranges from under one year in hyperscale centres to 5+ years in enterprise and colocation centres. Consequently, disruption to manufacture will influence current DC operators' ability to improve service (by upgrading equipment) or extend provision (by expanding current and/or equipping new data centres). Reliance on digital technology and services has grown to a point where interruption of many current services (e.g., health, traffic control) would have very adverse social and economic impacts; examples include increased costs which will eventually be passed on to end users and limit access to the internet by the poorest populations. Uncertainty about the supply chain will also affect longer term business and service planning.

The engineering profession is comprised of many specialisms (e.g., civil, mechanical, electrical and electronic, computing, aeronautical); the 'traditional' approach is mono-disciplinary, and expertise and knowledge tends to be specific and deep but narrow, and, consequently, problem solving has not been holistic in that challenges have been addressed by developing fixes for individual problems and linear thinking, rather than going back to first principles and/or adopting a whole systems approach. This approach is prevalent in the data centre industry, which is comprised of 10 internal subsectors and 1 external subsector (see Figure 1), each of which employs highly talented and skilled professionals. However, there has been very little interaction between the subsectors and, consequently, problems solved in one part of the system (subsector) have often created problems in other parts of the system, one example of which is described above, i.e., that of reducing energy consumption by increasing water use.

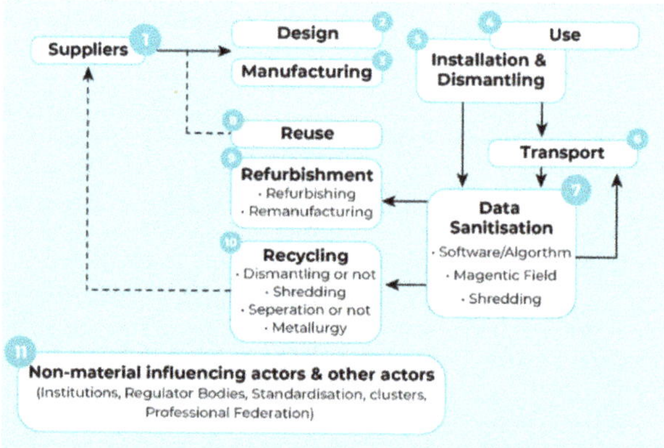

Figure 1. Data Centre Industry actors and sub-sectors, the CEDaCI project.

In summary, the data centre industry is unique: the rate and scale of growth of the technology and sector are unprecedented, as is the impact of services provided (specifically connectivity) on almost every aspect of daily life in the developed world, and influence is increasing in the developing world. To date, the sector has focused on the short term and provision of increasingly fast and extensive services, based on a linear take-make-use-dispose economy. The fundamental design of data centre equipment has not changed since the sector began, and most disruption and innovation are associated with the services provided by the technology, rather than the technology itself. The above factors necessitate an urgent and systemic review and change in practice across the sector to ensure that future services remain uninterrupted and that the sector becomes environmentally, socially, and economically sustainable. This could be assured through development of a sectoral Circular Economy although silo working and fragmentation currently presents considerable technical and behavioural challenges to this [21]. Section 3 will discuss CEDaCI–a project that was developed to initiate a Circular Economy for the Data Centre Industry; the methods employed to underpin this project are also described, and then, their impact is investigated, and assessed, via the quantitative and qualitative data in Section 4.

3. CEDaCI—A Circular Economy for the Data Centre Industry

The particular and diverse challenges of the data centre industry are unique and require an innovative whole systems approach to instigate major change because decisions made and actions taken at each life cycle stage affect the sustainability of all other life cycle stages. The Circular Economy for the Data Centre Industry (CEDaCI) project was initiated to kick start this change; it was developed and executed by academics and consultants from several different disciplines (including product and engineering design, social and behavioural sciences, business and life cycle management, and materials science), with knowledge of, but working outside, the sector, which is helping them to overcome these challenges. The project includes around 20 partners based in North West Europe; it is led by London South Bank University and other main partners include Operational Intelligence (UK), WeLOOP, TND and TEAM2 (France), GreenIT Amsterdam and SDIA (Netherlands), and Wuppertal Institute for Climate Environment and Energy (Germany). The project structure is underpinned by respected and proven methods for innovation, namely Design Thinking and the Design Council's Double Diamond process. Both methods also support a whole systems approach, which is critical to the development of any circular economy, and in particular, that within fragmented sectors.

3.1. *Method: Design Thinking and the Double Diamond Framework*

Design Thinking has always been integral to design practice and the design profession but the approach was introduced to the business community in 2008 [22] and wider community in 2009 [23] by Tim Brown, who defines it as 'a human-centred approach to innovation that draws from the designer's toolkit to integrate the needs of people, the possibilities of technology and the requirements for business success' (see Figure 2); it is also defined as a hands-on, user-centric ideology [24], and a non-linear iterative process that teams use to understand users, challenge assumptions, redefine problems, and create innovative solutions [25], and it has evolved to become an established practice. It has already proved successful to many sectors because it addresses the biases and behaviours that hamper innovation [26]; its value is also recognised well-beyond the design profession, and a number of reputed leadership and training establishments now offer courses in Design Thinking, such as Cambridge University [27] and Massachusetts Institute of Technology, for example [28].

Prior to Brown's public dissemination of Design Thinking, in 2005 the UK Design Council published the Double Diamond, which has become a universally accepted depiction of the design process (see Figure 3). The developers also acknowledged that the process and methods were already integral to, and widely practiced by, the design profession but the diagram was devised to structure the process more clearly and, therefore, to both support designers to progress, from problem to solution, and to help them to explain the process to clients. The original version lists four stages, the first two of which (Discover and Define) encourage broad and deep exploration of the challenge (divergent thinking), while the second two (Develop and Deliver) encourage focused action (convergent thinking) [29].

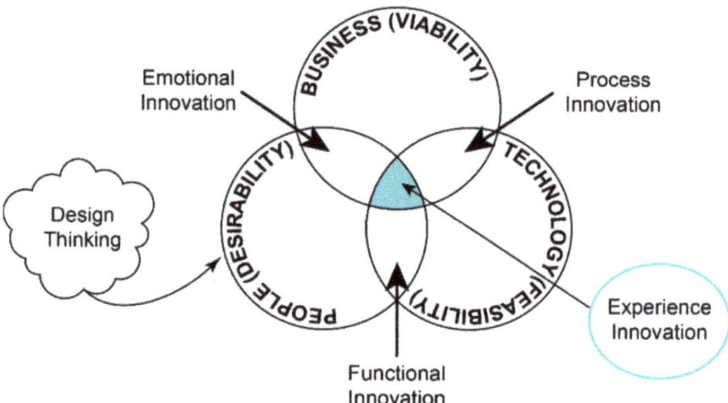

Figure 2. Design Thinking used to initiate innovation, adapted from IDEO.

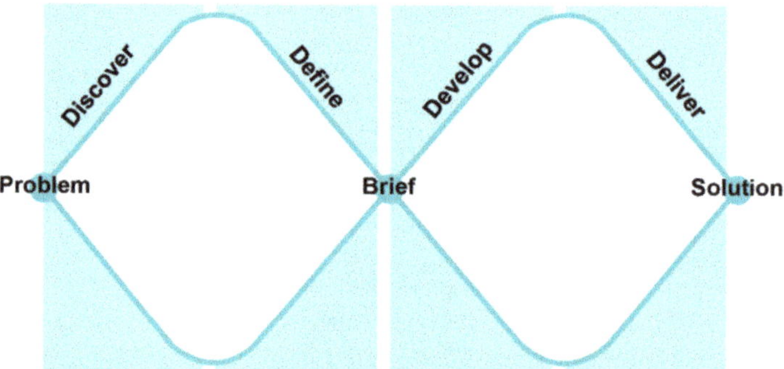

Figure 3. Adapted from Design Council Double Diamond, 2005.

The Double Diamond was also very successful within, and beyond, the design community, and in 2015, it was revised and enhanced to form a framework for innovation and to provide a structure for Design Thinking (see Figure 4). In Design Thinking and the Double Diamond framework, people (stakeholders) are critical to the process and successful project delivery. This is further clarified in the Double Diamond model, which includes stakeholders in all stages as follows. The entire process is underpinned by Engagement (building relationships between citizens, stakeholders, and partners) and Leadership (to create conditions to allow innovation, including culture change skills and mind sets). Design Principles and Methods were also added to the revised Double Diamond model and, again, both specify engagement strategies: Design Principles-1. be people centred; 2. communicate visually and inclusively; 3. collaborate and co-create and 4. iterate, iterate, iterate-and Methods Bank-1. Exploration (of challenges, needs, and opportunities), 2. Shape (prototypes, insights and visions) and 3. Build (ideas, plans, and expertise) [30]. The principles and methods are integrated throughout the process, so stakeholders should be regularly consulted to ensure that the project output meets their current and future requirements.

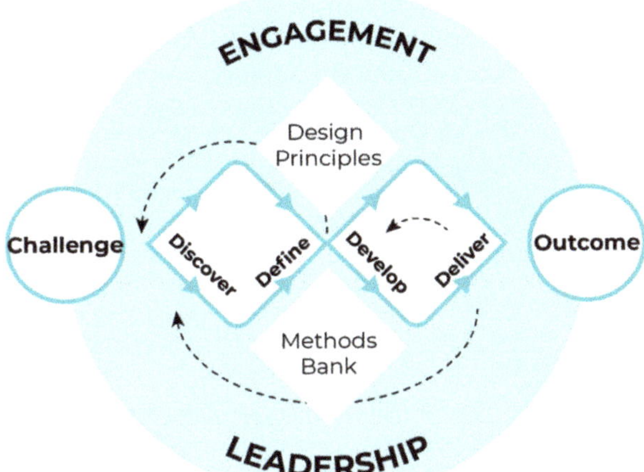

Figure 4. Adapted from Design Council Double Diamond framework for innovation, 2015.

3.2. Stakeholder Engagement and the CEDaCI Project

Stakeholder engagement has already proved beneficial to many sectors because it clarifies communication and the exchange of ideas and helps all parties to develop a thorough understanding of issues, alternative perspectives, and potential solutions; it also strengthens the resources of the involved individuals and groups by increasing awareness, confidence, skills, and co-operation, and it improves sustainability of the initiatives by increasing the quality of decisions and their acceptance among stakeholders [31]. The reasons for its success as a strategy are associated with motivation and fundamental human psychology: engagement has been defined as the sum of supportive conditions for authentic expression; it is motivated by either cognitive factors (such as a rational work goal), emotional factors (a state of mind that affects behaviour) [32], or motivated by a combination of cognition and emotions [33]. These drivers tend to vary according to role, so consumers are motivated by emotional factors such as empathy, gratitude, and trust [34] while stakeholders are motivated by more functional and rational (cognitive) factors [35], which may be evidenced as goal-directed behaviours such as accomplishing a predetermined purpose [36].

Stakeholder engagement was seen as critical to the CEDaCI project to enable and ensure a whole systems approach to the challenge, via knowledge exchange between the project team members and stakeholders and among stakeholders from the various sub-sectors. However, the history and evolution of and behaviours across the Data Centre Industry meant that there was a risk of non-engagement. The leadership team believed they could reduce this risk to a manageable level by stimulating interest through first- and second-hand contacts and by meeting potential participants in person at trade and other events, although this is a time and resource intensive activity.

The structure of the DCI means that members are often simultaneously stakeholders and consumers (i.e., they are involved in delivery and procurement and use of products and services) and therefore, they are motivated by both cognitive and emotional factors. In the context of CEDaCI, for example, as stakeholders, their eventual goal is to increase sustainability of the DCI and build a Circular Economy; as consumers, they meet the team partners and colleagues from across the industry, and by sharing experience and knowledge, they develop empathy and trust.

In order to reach and engage as many actors as possible during and after the project three types of activity were developed and employed:

- Working Groups (WGs): stakeholder-focussed meetings by invitation and referral where experts associated with the sector discuss and advise on general project activities to ensure they meet sectoral requirements; 20–25 participants per event
- Co-creation Workshops (CCWs): consumer and stakeholder focused events by invitation and referral where experts guide and review the design and development of specific outputs; 20–25 participants per event
- A virtual pan-sectoral Network (NM): consumer and stakeholder focused platform with open membership to publicise project activities, progress, news and events, and encourage knowledge sharing and connection among members during and after the project (100 members to date)

3.3. *Project Structure and Process*

As stated above the four stage Double Diamond framework project underpinned the project structure and stakeholders were also involved throughout every stage.

- *Discover*—research and analysis of the entire DCI and associated activities; secondary data collected from publications and primary data through contact with stakeholders
- *Define*—Brief, project scope and key outputs were developed in response to the above findings by the core CEDACI team and confirmed in Working Group meetings (with partners and stakeholders)
- *Develop*—regular WG meetings and CCWs with stakeholders to identify their requirements and wishes, share knowledge, and gain feedback to support iteration and execution of the Project Pilots (A—design and manufacture; B—product life extension increased by incentivising secondary market; C—new recycling processes and increased CRM reclamation) and digital tools.
- *Deliver*—Project output (e.g., bespoke Eco-design and CRM assessment tools; overall Data Centre sustainability assessment tool) tested and reviewed by stakeholders/consumers; tools and other output refined in response to feedback. Following final delivery CE training offered to 50+ SMEs (stakeholders/consumers).

The project and progress are continually publicised, via a dedicated website, and regularly promoted via panel discussions and presentations at global trade events (e.g., Data Centre World, Data Centre Dynamics), and specialist conferences (e.g., TechUK, Forum for Sustainability through Life Cycle Innovation) and via social media platforms to raise awareness, maintain a public profile, and engage with as many DCI and IT industry representatives as possible.

4. Results

The anecdotal response to invitations and attendance of the various meetings and events was positive; however, a more objective assessment of the level and impact of engagement and suggestions for improvement was sought in order validate the whole systems approach, the engagement strategy, and perceived impact of the project overall. Initially, quantitative data was collected via an online survey to produce a general overview of the response to the CEDaCI project and the level of stakeholder engagement. This was followed with a more comprehensive qualitative research using semi-structured interviews to gather deeper insights. (See Supplementary Materials for the list of questions).

4.1. *Quantitative Research*

The on-line survey of CEDaCI Network Members was published via the CEDaCI Network and comprised of 18 questions, most of which required multi-choice/tick box responses, but 3 also offered respondents the opportunity to add comments. Standard data collection and analysis software was used to set up and run the survey, which was live for 14 days. The response rate was fair (44/100 NM) and once the data were cleaned there were 32 usable datasets; although a larger dataset would have been better, it was reasonable for this type of survey and deemed valid during a global pandemic. COVID-19 adversely affected many businesses and individuals (e.g., through illness, work-life balance, caring and educator roles, furlough, and redundancy); however, there was a marked increase in data traffic and associated activity and workloads in the data centre industry as populations worked, studied, and communicated remotely, which limited time for DCI members to engage in non-essential activities like surveys and interviews.

The respondents included actors from 8/11 life cycle stages/subsectors as shown in Figure 1 above. Most respondents work in companies with more than one main activity while others work in companies that engage in main and secondary activities, as shown in Table 1 below, which also showed that the main businesses of respondents vary, and 8 are directly involved in data centre systems design and operations (2 and 4), 3 in refurbishment and reuse (8 and 9), 1 in de-installation (5) and 1 in recycling (10); secondary business activities include systems design (2), data sanitisation (7), refurbishment, and reuse (8 and 9) and recycling (10).

Table 1. Respondents to CEDaCI engagement survey—identified by subsector—32 responses.

Data Centre Industry Subsector	Main Business Activity	Secondary Business Activity
1. Suppliers	0	0
2. Design (DC systems)	8	2
3. Manufacturing		
4. Operation/use	8	
5. Installation and dismantling	1	
6. Transport		
7. Data sanitisation		7
8. Reuse	3	10
9. Refurbishment	3	10
10. Recycling	1	10
11. Other actors e.g., policy/regulation bodies		

Respondents' roles are as follows: the majority (18/32) are employed as data centre design, strategy, management, and operations consultants, or as sustainability and circular economy leads for global and national private companies, public, and non-profit organisations, and all respondents (32) hold senior positions, such as managers, senior managers, CEOs, directors, and owners.

Table 2 below shows that all 32 respondents were Network members by default but analysis of the results showed that 9 were only Network members, 16 participate in the Network and either the Working Groups, or Co-creation Workshops, and 7 participate in the Network, Working Group, and Co-creation Workshops.

Table 2. Respondents to CEDaCI engagement survey–level and type of activity–32 responses.

Activity	Number of Respondents
Network members	32
Network Members only	9
Network membership and Co-creation Workshops (CCW) or Working Group (WG) membership	16
Network membership and CCW and WG membership	7

The respondents have been engaging throughout the project and 27/32 volunteered information about the date that they joined the Network: 11 as soon as it was launched in January 2019, a further 10 during 2019 and 6 in 2020 Network membership is open to anyone whereas participation in the Working Group and Co-creation workshops was by invitation only or referral to ensure that there were representatives from as many subsectors as possible and numbers were manageable. The WG and CCW invitations were circulated soon after the project launch and again most respondents (16/19) joined, and have participated regularly, since invited, and 3 have joined between September and

December 2020 by referral, all of which indicates commitment to the project and activities (see Table 3).

Table 3. Respondents to CEDaCI engagement survey–length of membership–27 responses.

Joining Date-Network	Number of Respondents
January 2019-launch	11
During 2019	10
During 2020	6
Joining Date–CCW and WG	
January 2019–Launch	16
September–December 2020	3

23 Network members also connect with the CEDaCI team outside the organised events (see Table 4) but connection appears to be local (e.g., Network members based in the UK connect with LSBU, those in France with WeLOOP, in the Netherlands with GreenIT Amsterdam/SDIA, and with those in Germany with Wuppertal Institute); this could be due to language or mutual contacts in the local area, for example. Levels of contact also vary and although 9 members have not yet made personal contact 11 members make intermittent contact, 3 once a month, 6 once since joining, but 1 person contacts the project team at least once a week. Most respondents selected more than one reason for joining the Network: only 8 joined to provide advice and 13 to connect with other members of the Network; the most popular reasons cited were to share information (with the project partners) (17) and to follow project progress (17). With regard to the benefits of Network membership, once again, some respondents selected one benefit, others 2, and others 3. The main benefit was identified as being able to connect with colleagues in the same subsector (selected by 14 respondents), the second was to get information about data centre sustainability (9), and only 5 respondents saw connecting with other subsectors as beneficial; this is not entirely surprising and reflects the silo-working practice (see Table 4). As previously stated, silo working and attitude is common to the DCI, so the results were not entirely surprising and the project team realise they have to do more to engage cross-sectoral communication. Finally, 11 respondents follow CEDaCI on LinkedIn, 4 on Twitter, and 2 on both platforms, i.e., 18/32 respondents; this could be due to a low level of engagement with social media across the group or to ignorance about the CEDaCI accounts. The team will publicise the links more widely through other DCI accounts and work to increase Network membership and engagement.

Table 4. Respondents to CEDaCI engagement survey–reasons for joining and benefits: 23 responses.

Reason for Joining	Number of Respondents
To provide advice	8
Connect with Network members	13
Share information with project partners	17
Follow project progress	17
Perceived Benefits	
Connecting with individuals in same subsector	16
Connecting with individuals in different subsectors	5
Get information about DC sustainability	9

In summary, the respondents were very positive about the project and their experience; their early enrolment, and regular participation in activities, reflects their on-going commitment to the project.

They are also informed about sustainability, and their senior positions and roles, within their organisations, indicate that they all have potential to influence practice either inside and/or outside their organisation. They see participation as beneficial for networking, although connections tend to be local and limited to their sub-sector rather than cross-sectoral. There was no negative feedback, but there were several suggestions about increasing project visibility and marketing drives to increase uptake, including demonstration and training sessions for the UK government Sustainable Technology Group, which positively endorses the project and its output. This feedback was useful, but very general, and so, a further study was undertaken to gain deeper insights into the strengths and weaknesses of the project and stakeholder engagement so far.

4.2. *Qualitative Research*

In order to gain deeper understanding of the response to the project, and to find out whether the methods employed were successful or needed to be adjusted or revised, a more extensive survey was also undertaken. As stated above, DCI representatives were affected by the pandemic and although 12 candidates responded to interview requests and were invited to participate in the research activity, two then said they were unable to do so due to a high volume of work and one candidate had to withdraw at the last minute for personal reasons, so there were 9 participants. This response rate, and the results, are regarded as acceptable for the nature of the study.

Nine professionals associated with the data centre and digital technology sectors participated in semi-structured interviews, with 10 open questions, which lasted between 30 and 60 min. All interviews were conducted online using video/audio software and recorded; every recording was transcribed, then analysed, using specialist software, which enabled identification of particular themes and sub-themes; the results were also cross-referenced to identify trends in expertise, attitudes, and behaviours and to assess the success and impact of the CEDaCI project model and methods to date.

All nine participants hold senior positions within their organisations, as shown in Table 5, and include: 3 sustainability leads/managers in a refurbishing company (B), an assets disposition company (E), and a national non-profit organisation (C); a Circular Economy manager for a corporate IT producer (A); an associate director for climate, environment, and sustainability in a digital technology trade association (I); a vice president of a foundation for open source hardware and data centre design (G); an ambassador for a non-profit global ICT producer (D); expert in electrical and electronics materials (F); and a technical director of a data centre consultancy (H).

Table 5. Participants in CEDaCI Qualitative study –roles and employment.

Identifier	Participant's Role	Participant's Role
A	Corporate IT producer	Circular Economy manager
B	Refurbishment	Sustainability manager/lead
C	National non-profit organisation (digital section)	Sustainability manager/lead
D	Non-profit global ICT producer (open source hardware and data centre design)	Ambassador
E	Assets deposition	Sustainability manager/lead
F	University	Expert in electrical and electronics materials
G	Non-profit global ICT producer (open source hardware and data centre design)	Vice president
H	Data centre consultancy	Technical director
I	Digital technology trade association	Associate director for climate, environment and sustainability

4.2.1. General Contextual Questions, Responses and Evaluation

An overview of the responses to the general and individual questions posed is presented in Table 6 below; a more detailed review and analysis is also presented as follows: the first three questions were designed to confirm the participants' knowledge and experience in order to validate their responses to the specific questions about CEDaCI.

Table 6. Overview of responses to semi-structured interview questions.

Participants	A	B	C	D	E	F	G	H	I
Subject knowledge									
circularity		x	x						x
efficiency	x		x						
energy efficiency		x			x				x
materials efficiency									
economic		x							x
environment					x		x	x	x
social			x						x
conflict minerals									x
data security concerns			x						
Sustainability within company									
energy efficiency-now		x	x	x	x	x	x	x	x
economic factors-now		x	x	x	x	x	x	x	x
environmental factors-now		x	x	x	x	x	x	x	x
social factors- now		x	x	x	x	x	x	x	x
good practice in the future	x					x			
IT energy savings-necessary			x						
cycling-client visits etc.								x	
reduce impact of travel and transport		x						x	
sustainable practice in-house		x		x	x				x
philanthropic activity		x							
Project vision and change									
positive impact of CEDaCI	x	x	x	x	x	x	x		x
Awareness raising, contribution to developing CE and initiating change									
positive impact on Data Centre Industry		x	x	x	x	x	x		
Impact of CEDaCI on participants and employers									
beneficial support-now	x	x	x	x	x	x	x	x	x
brings specialist knowledge to company		x	x						
beneficial support-future	x	x	x	x	x	x	x	x	x
Value of stakeholder engagement									
new to CCWs and WGs	x		x	x		x			
prior experience of CCWs and WGs		x				x		x	x
cross-sectoral activities are beneficial		x	x	x	x	x	x	x	x
Future activities									
economics of CE			x						
ethical procurement			x						
broadening scope of LCAs etc.				x				x	
publicising output and educating DCI and other industries		x				x	x		

Participants were first asked for their definition of data centre sustainability: the most common subject was circularity, which was mentioned by (B), (C), (G), (H), and (I) who all had a good understanding of the principles of the circular economy, that the embodied impact of products is as important as that of operational energy and the need for a holistic approach to the challenge. (A)—the Circular Economy lead—did not mention circularity as such but alluded to it in points about the complexity of data centre sustainability, whole product life cycles, removal of redundant equipment, space used, and operational efficiency; both (A) and (C) commented on efficiency, and it is worth noting that both work for very large (global and national) organisations.

Economic, environmental, and social factors were each mentioned by 2 participants. (B) and (I) mentioned economics, and (E) and (I) mentioned the environment; all of their comments were linked to operational energy consumption, associated impacts, and use of renewables rather than physical resources, which was a little surprising considering the participants' comments regarding the circular economy. (C) and (I) mentioned social factors, although their emphasis was different, and while (C) referred to concerns about data security,

(I) referred to conflict minerals, which is keeping with their respective roles—(C) within a national operator that relies on DCI services and (I) a technology advisory organisation; in fact (I)'s entire response to data centre sustainability was the broadest, and all subjects, other than efficiency were specifically mentioned, which is also in keeping with role.

The second and third questions were about the participants' organisations (employers), what they are, and what they could be doing better in relation to sustainability. Economic, energy efficiency, social, and environmental factors were identified in response to question 2, and all participants other than (A) mentioned these in relation to current activities within their organisations, although (A) did talk about potential and future improvements. The subjects and responses were significantly interlinked, and participants referred to in-house and external practices (with clients and customers). All participants cited examples of good practice, but they varied according to the focus and size of the organisation and local political factors. For example, (H) works for a small consultancy, and all employees work remotely at home, or on site, with clients; although the employees are encouraged to cycle to sites, energy efficiency is dependent on personal habits at home. Conversely, (C) works for a huge national organisation and has driven good practice to make significant energy and resource savings across their IT activities. Employees of customer and client facing organisations ((B), (D), (E), (G), (H), (I)) are also driving good practice as part of their business and activities (e.g., increasing energy efficiency, extending product life, and reducing packaging and general waste to reduce environmental impact) and developing strategies, tools, products, and services to help their clients to do the same. There are also examples of philanthropy, and one refurbishment company donates unsalable, but fully functioning, products to organisations that develop IT skills in developing countries and support local charities that help people with special needs. There were several common observations, including the lack of robust recognised metrics and standards, for the secondary market and products; although several organisations are developing their own, they will not be comparable with each other, which may be confusing for customers who want to compare metrics.

Another common point relates to the increasing awareness of embodied impact and frustration at the lack of infrastructure, or policy, to guide and bring about significant change. (F) noted that, until recently, his organisation would not buy second life (refurbished) equipment because of issues around warranties; however, this is changing in response to policy change (specifically Wales' Well Being of Future Generation act), which is encouraging more circular practice, which could drive common metrics and standards. Finally, the general consensus of (C), (G), and (H) who are influencing change in house, and with external clients, is that sustainability has to be linked to the bottom line.

Unsurprisingly, subjects raised in response to question 3 included future, growth, and targets in addition to economic, environmental, and social factors. Both (E) and (G) were confident that their business practices are as sustainable as they can be at the moment, although (G) would like open-source hardware to be included in policy to increase market growth for this type of product as an alternative to being locked into single brands. This is contrary to (A), who works for a company that is seeking to increase brand loyalty by selling a service rather than equipment, one benefit of which is that the company will have control over the entire product life cycle from cradle to cradle. (B) and (H) mentioned that they could reduce the impact of travel and transport of goods, while (C) mentioned that the sustainability programme now includes resilience, so it can cope with emerging and future challenges, such as COVID-19, pandemics, and climate shocks. Finally, (I) commented that the organisation is based in rented premises, and they don't control waste and recycling at present, but they are measuring their current activities, so they can make their business activities more sustainable, and this includes signing up to the Race to Zero (carbon reduction) initiative. In general, all participants simultaneously recognised current good practice in their organisation and the need to improve.

This investigation and analysis of participants' current knowledge and workplace activities, practice, and attitudes provided extremely important insights into their business practice, role, and trends across the sector. It also confirms that their opinions about, and responses to, the CEDaCI project are based on high level of knowledge and experience. The subsequent questions related their reaction to, and perception of, the CEDaCI project as a whole and highlighted the value of stakeholder engagement.

4.2.2. Response to and Evaluation of the CEDaCI Project

Individual Questions

Question 4, the first in the section, relates to the project vision and asked whether CEDaCI is accelerating development of a sectoral Circular Economy: the response was almost unanimously positive and even the participant who responded negatively (H) qualified their response and commented that it was impossible *"because it is such a huge task but the project is raising awareness of the Circular Economy and is making decision makers think."* All other responses were very positive: and 4 further participants ((B), (D), (F), and (G)) commented that the project is raising awareness and *"has opened conversation that wasn't there before and people didn't bother thinking about before' 'despite limited resources."*

Awareness raising is the first step in the development of the Circular Economy; second steps involve supporting and empowering businesses and organisations to make change. The participants were again very encouraging and commented that *"It is the first* (project) *to investigate the challenge analytically"* and *"the potential is huge"* (B); *"the project is also bridging the gap between theory and practical guidance"* (C) and the output (specifically the digital Circular Data Centre Compass (CDCC)) *"is immensely valuable"* (F) and will *"support public and private organisations to make informed decisions about procurement and practice"* (E). Similarly, if we *"make those* (tools) *available to organisations and they are able to use them easily and adopt them, you've solved a huge issue that most places don't have the time to look at. If you simplify the process for them, it removes huge barriers"* (F).

Question 5 was designed to find out whether CEDaCI is supporting the DCI in general; all participants responded very positively, highlighted subjects were (again) awareness, circularity, green procurement, the public sector, market visibility, and the CDCC digital tool. The responses acknowledged the wider benefits and impact of CEDaCI; for example, *"The tools that you have provided will be hugely helpful for operators who currently still are very energy and carbon focused and they are not thinking about the embedded carbon in the assets that they use"* (I). The link between economics and sustainability was repeated and *"Anybody who adopts your tools and uses them properly should make more money. This means that they can save the planet and create social value and sustainable options for the future"* (F). The project output has the potential to influence change *"provided that important buyers like the government and others start requesting this information and demanding action it will drive the market accordingly"* (C). This could be realised because one participant is already referencing CEDaCI to government bodies, such as DEFRA, who are developing a Cloud Sustainability Standard and another wants to *"showcase the CDCC to the Cross-Government Technology Group to get feedback and as a game changer for their thinking"* (C).

Questions 6 and 7 were designed to gather more specific feedback about the impact of CEDaCI on the participants and their organisations now and in the future; some of the subjects identified were the same as those raised by other questions (business, future, and growth) but they also included planning and support. Again, all participants responded positively and commented that support for current business practice was highly beneficial; for example, *"Those involved in sustainability impact roles can use this information to make their case in their organisations for sustainable choices to be made"* (D). Examples of specific project-partnership tasks include checking other LCAs and carbon assessments in the public domain and creating new LCAs of open source hardware; participants (B) and (C) agreed that the CEDaCI team bring different specialist knowledge to their organisations and that external analysis and reports increase objectivity and credibility (*"The findings .... published by someone like you will be invaluable"* (G)). They also expect that project output will continue to support them in the future, as more output like the CDCC and Pilot projects are completed. Several participants also made suggestions for future work and collaboration after the project: for example, "*You have the data with the Compass and when we talk to the big systems integrator about the data centre contract, I think CEDaCI could help support us*" (E). Similarly, "*It's likely that sustainability principle will be added to our code of practice so you could look at how to mesh your thinking and tool into how the public sector operates*" (C).

Like question 4, question 8 was designed to learn whether the project process and output was fulfilling a key aim: as explained above, the DCI is fragmented and silo working is endemic across subsectors. Development of a Circular Economy requires a holistic approach and input from representatives from all life cycle stages across an entire industry. CEDaCI is seeking to do this, and although the (internal) team recognises that there is some progress and shortfalls, external feedback from participants is essential to confirm or negate this perception. Consequently, participants were asked whether the mix and composition of partners, Working Group, and Co-creation Workshop members is right. The question encouraged comments around the environment, suppliers, manufacturers, and users from 7/9 participants. The feedback was mixed, but this was expected because, despite major efforts to recruit DC operators, most said they were too busy to join the activities although consultants with prior experience of running data centres were a good substitute. A typical comment was "*There was a great mix of people. There could be more users involved in the development process*" (I). Similarly, the team were unable to engage manufacturers in the project regularly; they visited a global IT producer's site in Scotland, but their production plant, like those of other manufacturers, is outside the UK so the visit was to a "*technology renewal*" reuse and recycling centre. Again, low engagement was due to high volume of work and other typical comments were "*End users and manufacturers lacked representation, but that is understandable as many are distributed globally*" (H). The participants believe that there is potential to engage with more users and manufacturers when the outputs—and especially the CDCC tool—are complete.

Question 9 related to perception of the Working Groups and Co-creation Workshops and the value of stakeholder engagement: 8/9 participants responded when asked whether they had taken part in similar events prior to the CEDaCI events and 50% (A), (C), (D), (F) said no, or not in the way that the CEDaCI activities were run, and the other 50% (B), (G), (H), (I) said yes but made comments like "*I've learnt more from CEDaCI than from anywhere else apart from our own research*" (B). When asked about the benefits of the cross-sectoral WGs and CCWs to the CEDaCI project, and to them as individuals, all 8 participants responded very positively and the general consensus was that the events were very successful in initiating different ways of thinking about problems. For example, "*The key thing is a change of consciousness. You are getting people to think about new ideas*" (D). The format also encouraged open and objective conversations, which participants really appreciated: "*To get all those people involved was a major challenge, discussing openly problems and solutions. They are able to freely talk about ideas and mutual benefit*" (F); "*it helps to ensure that you can take account of scoping issues by bringing together the supply chain and actors that might not normally come together. It's good to hear different perspectives and have a balanced view, without it being influenced by vested interests*" (B). Participants also valued the interdisciplinary composition of the meetings and (I) "*found it interesting to hear other people's perceptions and views that I would not have had access to otherwise*" while (G) said "*each one of the meetings felt useful*" and "*the biggest benefit is meeting other people with other perspectives. The community aspect is one of the benefits of working groups*".

In the last question (10), participants were asked for suggestions about future activities or events: the response was encouraging and indicated that there is scope for future work to extend the impact of CEDaCI by organising a "*meeting or event where you discuss your approach to developing those tools and how you can help others to build similar things or transfer to different industries*" (F). Ethical procurement issues (C) and understanding "*more about the chemical processes that are used to extract elements*" (H) and (in response to HM Treasury Business Case) (C) said that factoring "*sustainability into business decisions and working that through to monetised values … would be really useful*"; broadening research and modelling to include all types of current and emerging DC equipment (D), were also highlighted as subjects for further investigation. The most significant suggestion that was made by several participants, concerning wider dissemination of project output to inform and educate the DCI, other industries, and the public: for example, (I) suggested a briefing event or blog for members of the technology trade association, (H) felt that publicising information about the overall manufacturing process would generate interest in the industry, (G) will be happy to do a joint webinar for their business community so they can see the end results once the tool is released, and (B) said *"you can reach a lot wider audience and you can educate them about what's in the technology and from a sustainability point of view, it has the benefit of educating people about data and ICT as well."*

Aggregated Data and Trends

The key themes identified in the transcripts are shown in Figure 5 below, which also shows frequency. Aggregation and analysis of qualitative data, generated by responses to all 10 questions, showed that of the 17 themes identified in the transcripts the most common was the environment (identified 15 times), followed by circular/circularity and economic (8); efficiency/energy efficiency (7), social and support (6), awareness, future, manufacture, and public sector (4), growth and market visibility (3), business, green procurement, target, and users (2) and tool (1). Considering the interests (i.e., all bar one have engaged with the CEDaCI project and take part in WGs and CCWs), expertise and roles within the organisations (5 are directly linked to sustainability, 3 are linked to overall efficiency and 1, sustainability-linked research) these results are not surprising; similarly, alignment of the themes with the individual participants reflected the type of organisation with which they work: for example the (A) (who works for a corporate global IT producer) focussed on efficiency, growth, manufacture, and targets; (B) and (E), who work for SMEs specialising in refurbishment/second life products and recycling, focused on the three tenets of sustainability, education, and circularity, as did (C), who also commented on the public sector and works for a national non-profit organisation; (D) and (G) work for a non-profit global open-source IT provider, and both referred to sustainability-linked criteria and the future growth; (F) (who works in a university) focused on the environment and economics as did (H), the technical consultant who also spoke of manufacture and business. Finally, (I) who works for an advisory body, mentioned the broadest range of themes, including sustainability and circularity, business, manufacture, planning, public sector, the future, and targets.

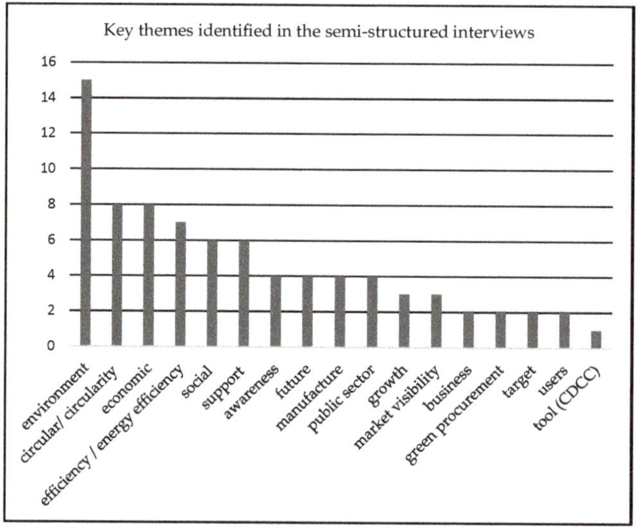

Figure 5. 17 Key themes identified in the semi-structured interviews by frequency.

MDPI
St. Alban-Anlage 66
4052 Basel
Switzerland
Tel. +41 61 683 77 34
Fax +41 61 302 89 18
www.mdpi.com

Sustainability Editorial Office
E-mail: sustainability@mdpi.com
www.mdpi.com/journal/sustainability

www.ingramcontent.com/pod-product-compliance
Lightning Source LLC
LaVergne TN
LVHW070704170726
843515LV00004B/485

* 9 7 8 3 0 3 6 5 1 5 3 2 8 *